Built by Animals

To Graeme

Built by Animals

The natural history of animal architecture

MIKE HANSELL

OXFORD
UNIVERSITY PRESS

OXFORD
UNIVERSITY PRESS

Great Clarendon Street, Oxford OX2 6DP

Oxford University Press is a department of the University of Oxford.
It furthers the University's objective of excellence in research, scholarship,
and education by publishing worldwide in

Oxford New York

Auckland Cape Town Dar es Salaam Hong Kong Karachi
Kuala Lumpur Madrid Melbourne Mexico City Nairobi
New Delhi Shanghai Taipei Toronto

With offices in

Argentina Austria Brazil Chile Czech Republic France Greece
Guatemala Hungary Italy Japan Poland Portugal Singapore
South Korea Switzerland Thailand Turkey Ukraine Vietnam

Oxford is a registered trademark of Oxford University Press
in the UK and in certain other countries

Published in the United States
by Oxford University Press Inc., New York

First published 2007

British Library Cataloguing in Publication Data
Data available

Library of Congress Cataloging in Publication Data
Data available

Typeset by SPI Publisher Services, Pondicherry, India
Printed in Great Britain
on acid-free paper by
Biddles Ltd., King's Lynn, Norfolk

ISBN 978–0–19–920556–1

1 3 5 7 9 10 8 6 4 2

Contents

List of Figures

Preface

It is a family story, and therefore probably untrue, that when at the age of about six I went missing on my short walk to school, I was eventually found watching a beetle negotiating the gutter. My first natural history enthusiasm was certainly insects. Later, an undergraduate project on the case-building behaviour of caddisfly larvae led to a doctoral thesis on the same subject. By that time I knew more than most people would wish to know about caddis larval cases. Leaving them completely behind, I spent two years teaching at the University of Khartoum in the Sudan, returning to the UK as an Assistant Lecturer at the University of Glasgow in 1968. On arriving there I was asked to give three lectures that probably changed the direction of my career.

At that time, Glasgow final-year Zoology students were given, apart from their special subject lectures, a series of occasional lectures called the 'A Course'. This in essence consisted of a handful of us academics each giving a few lectures on whatever took our fancy. Three lectures on caddis cases seemed excessive, so I decided to go to the other extreme and talk about everything that was built by whatever animal. This, I decided, was not to be a mere parade through the animal kingdom but an attempt to make some sense of, and discover some pattern in, the behaviour of that miscellaneous bunch of web spinners, house, case, mound, and nest builders. That is what I have tried to do ever since, and this book is an attempt to explain that biology, the biology of animal architecture and building behaviour, to anyone to whom that sounds appealing. It is a book about who builds, how they build and what those buildings do. It touches upon environmental impact, animal intelligence, architecture, engineering and building materials, on the organization of workforces, on traps, tools and art.

Writing this book has not been the lonely experience that some authors complain about. I would like to thank all those colleagues and friends who, through their daily interest and encouragement, contributed to its completion. This is my first experience of writing for a non-specialist audience, so I am very grateful to Lorna and Rowland Mitchell who, in the role of general readers, gave me valuable feedback, and to Jacquie Marshall who read and tactfully commented on the whole text. For specialist biological information and advice, I would like to thank Martin Burns, Robin Dunbar, Geoff Hancock, Felicity Huntingford, Bob Jeanne, Bill McGrew, Aubrey Manning, Maggie Reilly, John Riddell, Flavio Roces, Douglas Russell and Richard Wrangham. For very helpful information on human architecture, I must thank Jonathan Hale. For an enjoyable adventure to find the church of All Saints, Little Shelford, my thanks to Martha Jennings. Finally, I want to express my enormous appreciation to Graeme Ruxton. His broad interests, his enthusiasm, and his readiness to sit down and discuss points of biology have contributed immeasurably to the development of the final text. He was not only prepared to read everything but also read again my revised drafts. I dedicate this book to him.

1

The Builders

Standing with its back to the Atlantic atop a 90m (nearly 300ft) vertical cliff, at the base of which the ocean seethes and foams, is the great semi-circular, prehistoric fort of Dun Aengus. The location is the very westernmost fringe of Europe, the Aran Islands off the coast of Ireland. On the landward side, the fort's protection from attack is man-made. The half-moon inner sanctuary of the fort (45m or 148ft across) is embraced by a drystone rampart reaching over 5m high and nearly 4m thick. Beyond this are a further three irregular semi-circles of defensive walls and, should even the outer one be overrun, the attackers must then negotiate a field of massive limestone shards, jagged and stood on end. This is a *chevaux-de-frise*, a term also used to describe the barbed-wire enlargements that protected First World War trenches, but literally 'horses of Friesland', spiked obstacles used at least from the seventeenth century to break up cavalry charges. The age of Dun Aengus is uncertain, although it is probably over 2,000 years old: more surprisingly, we still do not know what threat could have stimulated the building of such formidable defences. Our need for protection is not any different from that of other species. Animals have always had enemies, in particular the climate, and other species seeking to prey upon them. We are just a newcomer species resorting to building to protect ourselves in a threatening world which includes our fellow humans among our enemies, and we are not the first species to make use of a *chevaux-de-frise.*

Think of the dangers that beset insects in a tropical environment. Probably the two greatest are predation by birds and by ants. The hairiness of some caterpillars is a defence against one or both of these, but ultimately the caterpillar must become an adult butterfly or moth. This requires a radical reorganization in its body design—crawling around eating leaves is replaced by flight, nectar feeding and mating. This transition can only be accomplished as an immobile, and therefore vulnerable, pupa. The commercial silkworm typifies the defensive strategy of moth caterpillars. It spins a silken cocoon to enclose itself completely, then it casts off its caterpillar skin to become a pupa from which a moth will later hatch. The silkworm is a hairless caterpillar, but imagine the consequences of a hairy caterpillar pupating inside its cocoon. Not only will it be shedding a hairy skin that crowds the inside of the cocoon, it will also be casting away protection that has already served it well against the threat of ants and even birds.

The caterpillar of the moth species *Aethria carnicauda*, is densely covered in long hairs. When ready to pupate in its native habitat of Central American forests, it selects a straight plant stem as a site to prepare its pupal defences. Facing down the stem, the caterpillar reaches back over its body to pluck out its hairs one by one with its jaws. Each hair in turn is secured to the stem, using the silk thread which is extruded from its mouth. In this way, the caterpillar builds up a disc of radiating bristles, a barrier to anything trying to pass up the stem. On completing this, the caterpillar backs up the stem a short distance and pulls out another series of hairs to create a second barrier and after that a third, even a fourth. It now turns round and moves up the stem a few body lengths and attaches another whorl of hairs above itself, backs down the stem a bit before plucking out more of its hairs to complete one or more additional barriers. The caterpillar is now secured by multiple lines of defence from ant attack directed up or down the stem, but it still has some body hairs left.

These are plucked out as the cocoon is spun and incorporated into its wall as a last-ditch defence against ants or other predators. Inside the hairy cocoon, the caterpillar now sheds its denuded skin to become an immobile but well-protected pupa.

A *chevaux-de-frise*, which features so impressively in the fort at Dun Aengus, is an ingenious defensive concept, so it is a fascination to us that a mere caterpillar is able to build a comparable device. We, as the world's pre-eminent builders, can't help admiring such skill and ingenuity. But, if asked to explain the basis of our admiration, we are confused. Is it because such a simple creature is able to make something so complicated, or is it that the animal has revealed itself as not so simple as we had thought? What goes on in the brain of the caterpillar when it is building? Is it anything like the way that we think when creating a similar structure? That is, does it have a plan? If not, and the construction sequence is a simple 'mindless' programme of assembly, how can the outcome be so sophisticated? Our delight and our confusion are essential themes of this book, a book that is about animal architecture—a subject that, before going any further, I need to define.

The Great Barrier Reef is not so much a wall of coral as a string of thousands of separate reefs that stretch 2,000km along the north-eastern margin of Australia to touch the southern shores of New Guinea. It has been created from the calcium-based secretions of coral polyps, colonial creatures closely related to sea anemones and indeed jellyfish, colonies of different species creating different shapes, 'stag-horn' and 'brain' corals, names illustrating some of the variation in shape that they show. I've heard the Great Barrier Reef described as 'the largest structure on the planet built by living things'. I am happy to agree with that, certainly for non-human structures. But I am not going to discuss coral reefs any further within these pages. The reason for this is that we cannot usefully ask of coral polyps what goes on in their brains—or, rather, in their nervous systems since they have no obvious brain—when they are creating their reef. It is not informative to ask this question in the same way that it tells me nothing about how your toenails grow to ask you what you think about when you are growing them. Coral polyps just secrete coral skeleton, gradually building up reefs. The caterpillar building its pupal defences on the other hand employs behaviour. In this book I focus on building that requires behaviour. This is no arbitrary demarcation because to have behaviour an animal must move different parts of its body in a coordinated way. This in turn necessitates instructions based on decisions. Nervous systems or brains make decisions, even if they are very simple ones, and nerves carry the instructions as nerve impulses to activate muscles to create movements.

My use of the word 'decision' in this context needs some clarification. I certainly don't mean to imply that the decision is a conscious process, and so the mind of a caterpillar is like our own. I mean something that could be a lot more simple. At its simplest, I mean the point at which information flowing along a single nerve A, comes to a point of contact with two other nerve fibres (B and C). Only one of these will be activated. If the signal in A is weak it will be B, if the signal in A is strong it will be C. This is a point of assessment of the signal in A; the point at which a *decision* is made. This is a familiar usage of the word in the context of information technology. Of course, it is true of essentially all the species of builders in this book that we have no idea what nerves perform what functions when the animal is building. Instead we use the powerful 'black box' technique to investigate their decision-making process.

The principle of the black box technique is to treat the animal as a closed box containing a mechanism which we cannot examine directly but which we can gradually come to understand by seeing how it answers certain questions. That probably sounds a bit abstract, but consider the giant golden digger wasp. The female of this handsome insect digs a short burrow angled into the ground, at the end of which she hollows out a horizontal, rounded chamber. Let's just ask the wasp one simple question about this sequence: 'when does she stop digging the tunnel and start building the chamber?' I say she stops digging when she feels that she has done enough work (i.e. by checking her internal state), but you say she measures the length of her burrow (say, by the number of steps it takes her to get to the bottom). We test which one of us is right by playing an experimental trick on the wasp as she interrupts her bouts of digging. We will do a little bit of the digging ourselves when she is temporarily away (feeding perhaps). My explanation predicts that she will do her normal amount of work and so end up with a burrow that is too long (her bit plus our bit). You are predicting that her burrow will be the normal length, and we will have just saved her a bit of digging. As it turns out, you are right. We now know how one type of decision is made within the black box of the wasp's mind. By further experimental interrogation of this kind, we can fill in further details. I hope that explains why from this point I'm going to ignore the corals of the Great Barrier Reef, and focus on building behaviour. We want to explore the complexity of decision-making processes associated with building behaviour.

I should point out in passing that you and I have just done something fundamental to science. We have conducted an *experiment*. We have observed the behaviour of an animal and, on the basis of that evidence, come up with rival explanations or *hypotheses* about its decision-making. These explanations have the crucial attribute that they make different *predictions* of what the animal will do when placed in a particular situation. We create that situation and see which hypothesis can be rejected, and which (if either) will be supported. Hypothesis testing is a crucial tool in science. This book will repeatedly refer to it.

To create buildings, another essential ingredient has to be combined with the behaviour—the materials. These may be collected from an animal's surroundings or secreted by the animal itself. Either way, according to the definition of building used here, behaviour must be used to create the structure. What the silkworm secretes is a continuous fine thread; only by shifting its body about does it build a structure (the cocoon) from this raw material. Birds nearly always use collected materials, grass or moss for example, to make their nests.

A theme that runs through this book is how behaviour is applied to different materials to make structures.

Now for a couple of housing stories. Standing on stilts in the waters of Loch Tay in Scotland is a round wooden-framed family house with a conical thatched roof, its only link to the shore being a narrow raised walkway. The vertical poles that support the house are embedded not in the mud of the loch floor but in an artificial rubble mound. This house is a reconstruction based on local archaeology of a *crannog*, a type of lake dwelling built in Scotland and Ireland around 2,000 years ago, the remains of which can be detected today as low circular mounds of rocks standing in shallow water. Building a house like this was clearly much more work than building a home on solid ground by the water's edge. The extra effort required to build a crannog must have been worthwhile because of the security offered by the encircling water. The same concept is embodied in the design of the lodge of the North American beaver (*Castor canadensis*). The lodge is a family home in the centre of a pile of branches rising above the water of a small lake. Beavers, being excellent swimmers, do not need a walkway to lead them from the shore to their front door. In fact the entrance to the house is underwater and so not apparent to a passing predator such as a wolf. The only significant connection between the single dwelling chamber and the outside air is through the upper part of the roof where the branches are packed more loosely, allowing fresh air to percolate down to the beavers within.

This lake house is, however, not the limit of the achievements of these beavers because the water in which it stands is frequently an artificial lagoon resulting from the construction of a dam created by the beavers themselves across what may initially have been quite a small stream. This dam is no randomly arranged barrage of woody debris either, but built with design elements that we can clearly recognize in our own engineering.

At the start of building their dam, beavers place sticks and branches where rocks or boulders are already obstructing the water flow. From this foundation, a low barrier is extended across the whole width of the stream causing a pool to back up behind it. Branches are then dragged through the water and over the top of the dam to be pinned into the stream bed on the downstream side or angled against the dam to form buttresses, with one end in the stream bed and the other against the dam wall. These supports reinforce the dam against the growing pressure of water in the lagoon. On the upstream side, not only are more branches added, increasing the thickness and height of the dam, but boulders from the stream bed are pushed up against the base of the dam while mud and fine debris are used to seal the wall. In this way, the height and thickness of the dam increases, but at all stages the top of the dam remains more or less level ensuring that no building effort is wasted. A completed dam may easily extend 50m (164ft) in length and exceptionally up to 200m, and be 5m in height.

Even this is not quite the extent of the building achievements of beavers, because the lagoon does not simply represent a broad defensive moat but also a larder. Beavers eat tree bark. During the summer they cut down sizeable trees, dismember them and drag the branches into the lagoon through specially constructed canals that radiate from it; there the branches gradually waterlog and sink. In the winter, when the lagoon is frozen over, the beaver family is still secure in its thick walled lodge and, should any of them feel hungry, they only need to dive out through the doorway and under the ice, across which wolves may roam, to retrieve a few branches to chew on.

Well, isn't that just wonderful! But what is the point of the beaver story? I have over the years been involved in radio and television programmes that celebrate animal builders. I'm very happy with '*Isn't nature wonderful!*' programmes because to watch animals build, or fly or perform courtship displays is indeed a delight and a wonder. But in this book we are not allowed to have it so easy. 'Isn't nature wonderful?' is now a serious question, and we need to do some serious wondering. So it is essential to confront what is, in this context, a problem. It is our emotional attachment to the living world and in particular our feelings towards animal builders.

Humans love music. This love supports a multi-billion pound, worldwide industry. I asked two friends, who are great opera enthusiasts, for their choice of most wonderful operatic experience. After deliberation, they decided upon the aria 'Casta Diva' from Bellini's *Norma* sung, in a live performance they had attended, by Jane Eaglen. My musical favourite would be the voice of the gospel singer, Mahalia Jackson, but I also experience a thrill at the distant conversation of a flock of geese heralding the approach of winter. To us these are wonderful sounds with evocations that are impossible to articulate. Animal-built structures also have an element of that for me. The nest of a long-tailed tit is one of the most intricate of any British bird. This neat rounded ball is dappled on the outside with a mosaic of pale lichens and flecked with a sprinkling of white spider cocoons. The glimpse of soft feathers seen through the small circular entrance conjures up feelings of comfort and intimacy, an emotional response to the natural world that I suppose I could call joy. Nothing wrong with that, unless it begins to cloud scientific objectivity.

There is a second obstruction to our rational judgement of animals, one that specifically besets scientists in the field of animal behaviour: *anthropomorphism*. This is the tendency for us to attribute human aims, thoughts and feelings to other animals without proper evidence. I was warned as a student that anthropomorphism was a sin against science. In 1872 that brilliant and versatile scientist Charles Darwin published his book *The Expression of Emotions in Man and Animals*. This includes, among many skilful etchings by a certain T. W. Wood, the face of a chimpanzee bearing the caption 'Chimpanzee disappointed and sulky'. Soon after this, it became impossible for a scientist in the field of animal behaviour to say such a thing and still retain their scientific credibility, and remained so for the best part of a hundred years. The reason was that by the start of the twentieth century, scientists were frustrated by fruitless debates on the relationship between mind and body generated by the approach of so-called psychological introspection (what you might call 'thinking about thinking'). Scientists in the then young discipline of animal behaviour, keen to stress their experimental approach and objectivity, rejected this method. Obtaining data on whether or not chimpanzees went into sulks seemed unrealistic, and speculating about it in the absence of such data, pointless.

I have to admit to always rather liking anthropomorphism not, I must emphasize, as an explanation but as a source of ideas or, as I should say, of hypotheses for stimulating lines of investigation. It was therefore a landmark for me and for the science of animal behaviour when in 1976 Donald Griffin, at the end of a distinguished scientific career studying the mechanism of echolocation in bats, published a book of a very different character entitled *The Question of Animal Awareness*.[1] It was an announcement to animal behaviour researchers that their discipline was now mature enough and should be confident enough to investigate, through experiment and hypothesis testing, the nature of animal minds. The chimpanzee might indeed be disappointed, even sulky. These were questions which we could admit to being interested in, and which we should try to find methods to study. While writing the first draft of this chapter in late 2005, I saw an obituary to Donald Griffin written in an august scientific journal, actually entitled 'Thinking about Thinking'. The message for the study of animal building behaviour is therefore that no question about what animals think or feel when building should be considered out of bounds to scientific enquiry. As a consequence I shall try in this book to push explanation to the limit and speculate on how our understanding could be extended, at the same time admitting candidly the extent of our ignorance.

Anthropomorphism can be a sin, or perhaps I should say a vice, because it is so easy for us to slip into it as a form of explanation rather than of speculation. When we see a beaver in a wildlife programme pushing a stake into the bed of a stream or positioning a branch to support its dam, we admire it for using engineering principles that we also use. In a similar way, we identify with the caterpillar for its ingenuity in creating a *chevaux-de-frise* from its own body hairs. In making use of these construction 'ideas' beavers become 'like us'.

We are special animals and one thing that confirms our belief that this is so, is the extent of our accomplishments as builders. It is interesting to look at the landscape in the Steven Spielberg film *Jurassic Park*. We see dinosaurs ambling or galloping through it, but it is easy enough in one's mind to substitute this scene of over 100 million years ago with one of only three million years ago when (after the extinction of the dinosaurs) the land was populated by large mammals resembling elephants and antelopes, all doing their own ambling and galloping. But look at our landscape now with its skyscrapers, shopping malls and motorways; how different it is from the one that prevailed for the previous 100 million years and more. We are, if nothing else, builders, and that leads us to admire other builders.

This has been something of a diversion from investigating builders and what they do, but I think that it is essential to bear in mind from the start the influence of these two powerful forces that can distort our understanding of animal builders. The first, a general emotional attachment to the living world; the second, anthropomorphism, which, in this case, is a feeling that animal builders are somehow special because they share some of the attributes that make us think of ourselves as special.

There is a third distortion to the mirror of our judgement: builders leave the products of their behaviour. This claim may seem a bit perverse since, having the tangible outcome of an animal's behaviour, a hairy caterpillar cocoon for example, to study at leisure separate from the animal itself is clearly very helpful. The problem is that, by comparison, it becomes easy for us to underestimate the skills and cognitive abilities that non-building species exhibit on a daily basis. So here are a couple of examples of behaviour which leaves no material evidence of its complexity yet, on careful observation shows sophistication. One concerns object manipulation by an ape, the other cognition in a bird.

Mountain gorillas are vegetarians but, in spite of first appearances, a mountain gorilla doesn't just tear a leaf off a nearby plant, jam it in its mouth and chew it. Leafy plants generally and unsurprisingly have evolved features that make them unattractive to leaf eaters, necessitating specialized feeding behaviour. One such species is a thistle that forms part of the diet of mountain gorillas. The leaf margins of this thistle are beset with spines as are the ridges that run up the main stem. The obvious discomfort shown by juvenile gorillas when eating these thistles demonstrates that feeding on them successfully requires experience. However, adult gorillas handle the thistles with confidence by holding the stem, orienting the spines, and folding the leaves in ways that effectively neutralize the thistle's defences, before the leaves enter the mouth. To become skilled in processing this and other food plants requires the gorillas to learn quite complex manipulation routines employing both hands. A detailed study has revealed 222 different handling behaviours. But, since many of these are minor variants, it has been possible to reduce this list to 46 functionally distinct elements; however, these are not employed randomly but organized into a total of 256 recognizable handling techniques or sequences of elements.[2] This remarkable dexterity goes largely unrecognized because the carefully prepared bundles of thistle are ground to a pulp and swallowed. For me to make a paper aeroplane probably requires less complex processing. If gorillas made paper aeroplanes rather than food bundles, then every museum would have one and every schoolchild would know about them.

Another feeding example illustrates reasoning processes in animals; it comes from a North American bird, the Western scrub jay. Like a number of corvids (birds of the crow family) it hides or caches food items to recover and eat at some later time. It is also a group-foraging species so that, when a bird hides a piece of food, there is a danger that this is noticed by a flockmate who later steals it. Field observations reveal that such theft commonly occurs, and aviary experiments show that a bird observing another caching food is quite skilful at relocating it. To counter this kind of theft, birds that know they have been observed when initially caching, later move food items to another location when there is no other bird present. This is in itself very interesting, but young naïve birds don't in general exhibit this re-caching behaviour. However, they begin to do it after the first time that they have stolen from a cache that they saw another bird make. Having been a thief, they apparently come to appreciate the possibility of being stolen from. This level of cognitive sophistication is almost certainly greater than anything that a termite can muster, yet we celebrate the marvels of termite mound architecture. What are we celebrating? That animal builders leave a permanent record of their behaviour is certainly an advantage to scientists but, in assessing the significance of their behaviour, we must be careful to make fair comparisons with the behaviour of species that leave no such record.

Three million years ago, let's imagine a Martian space-traveller meeting up with a Venusian time-traveller over a glass of something in a hyper-space bar. They fall to talking about Earth, which the Venusian has recently visited in his time-travelling. 'Guess what,' he says, 'in three million years time they will have quite an advanced technology, with air travel and the beginnings of space travel.'

'Surely not,' says the Martian. 'I was on Earth a couple of years ago and saw scarcely a sign of technology just an assortment of creatures making primitive shelters. Where do their engineers and technologists come from?'

'Would you believe it,' says the Venusian 'from the apes!'

'What, from that lot? You're kidding me. They don't build anything. I have had a stick waved at me once or twice and I've heard they can shape stones a bit but I would have put my money on the birds. I brought back a nice bird nest from Earth last time. I have it on my mantelpiece. Clever craftsmanship using several materials.'

This same point is made compellingly in the opening sequence of Stanley Kubrick's film of Arthur C. Clarke's *2001: A Space Odyssey*, where a looming, angular, alien monolith presides over the discovery by our ape-like ancestors that a bone can be broken by striking it with another bone. How else could a bunch of ham-fisted, hairy Earth creatures have done it, if it wasn't with some outside help? We need to try amongst other things in this book to answer that question, but first we need to review the whole range of builders that currently inhabit this planet.

The distribution of builders throughout the animal kingdom is apparently haphazard, so it is first necessary to recognize this before attempting to understand it. 'Kingdom' in this context is no mere poetic flourish, but a technical term in biological classification. Traditionally, the highest level of classification in the hierarchy of living things has been a *kingdom*. There are recent variations on or alternatives to this, but separating all living things into five kingdoms is still a simple way of recognizing fundamental distinctions. Under this system, the plants, fungi and animals are all separate kingdoms. The other two kingdoms are the protista, and the bacteria. The former is a miscellaneous collection of mostly single celled, simple organisms, which includes the amoeba. The latter is self-explanatory, except to say that around thirty years ago biologists began to discover minute organisms living in extreme environments such as volcanic hot springs, that superficially resembled bacteria. They are now allocated a group of their own, the archaea, with roughly the equivalent status of a kingdom, so you might say that we now have six kingdoms. That is by the by since the archaea have no relevance to the theme of this book. However, their ability to survive in extreme and unusual environments has enlivened the debate on what planets or moons might be capable of supporting life.

The next level down the hierarchy of classification is a *phylum*. All the vertebrate animals belong to a single phylum, the chordates, which contains around 100,000 species. Most chordates have a jointed, bony support along their back (i.e. are true vertebrates), but some possess a less well-developed skeletal support. The invertebrate animals, by contrast, are divided into over thirty phyla. Together, they have been estimated to contain between 10 and 300 million species. These numbers may sound so vague as to be largely guesswork, which is not far from the truth. What we can be sure of is that our estimate of the number of vertebrate species is quite accurate, while that of the invertebrates is not. We do continue to find the odd new species of mammal and bird almost every year, but they are essentially all known to science; even for the fish species it is estimated that over 90 per cent have already been discovered. By contrast, the invertebrate phylum the arthropods, which includes the insects, crustacea (crabs and relatives) and spiders, is estimated to have a massive 85 per cent of its species still unrecognized and unnamed by science.

Moving down the hierarchy of biological classification again, the next level is a *class*. The mammals are a class of vertebrate; the birds also are a class or, according to some more recent schemes, it is the group of repto-birds (Reptiliomorpha) that are more properly a class, with the birds being a bunch of feathered dinosaurs that have so far evaded extinction.

Beavers, chimpanzees, ourselves: we are all mammals, so called because all infants of this group are fed initially on mother's milk, which is secreted by mammary glands. That is of no significance here other than to say that mammals share this feature for the very important reason that, through evolution, they all diverged from a common ancestor that had this character. So, as we nearly all have come to accept, chimpanzees are our relatives as, more distantly, we are also related to Brants' whistling rat (*Parotomys brantsii*) and, more distantly still, to the Australian marsupial mammal, the hairy-nosed wombat (*Lasiorhinus latifrons*). 'More distantly' in this sense refers to how far in the past we shared a common ancestor with these kinds of mammals. The divergence of the marsupial ('pouched') mammals and the eutherian mammals (those like us that develop in a true placenta) seems to have occurred about 60 million years ago, shortly after the great extinction of the dinosaurs. This has given us modern marsupials, such as kangaroos, the koala and the hairy-nosed wombat, and modern eutherians that include the North American beaver, Brants' whistling rat and ourselves.

Below the level of class in the scheme of classification, we get *order*. Humans belong to the order of primates, which includes the monkeys and apes. The first primates appear in the fossil record about 55 million years ago. The primates are divided into a number of *families*, the names of which don't concern us here. Humans, we know from pub quizzes or just ironic observation, are referred to scientifically as *Homo sapiens* (literally 'wise man'). These two words define us according to our genus (*Homo*) and to our species (*sapiens*), the lowest two levels of the classification hierarchy. There have been other species of *Homo* before us, most recently *Homo erectus* ('upright man'). Others may follow us.

Beavers, as we have seen, are rather expert builders and so prime candidates for inclusion in any TV wildlife programme on animal architects, but what about the rest of their order, the rodents? There are rather a lot of species of rodents, and they make up a large proportion of the modest total of all mammal species. More than one in three, that is 1,500 out of 4,000 species of mammals, are in fact rodents. But what rodent other than the beaver should be included in any animal builders TV programme? Not a lot else springs immediately to mind. Nevertheless there are some competent if less celebrated rodent nest builders. The harvest mouse (*Micromys minutus*), widespread in Western Europe, builds a hollow ball of interwoven grasses typically supported by the stems of corn. In Africa, the tree rat (*Thallomys paedulcus*) builds an irregular nest of twigs and grasses in the branches of acacia thorn bushes. In Australia, the greater stick-nest rat (*Leporillus conditor*), a creature of about 350g (12oz), builds a family nest that may become a metre high and one and a half metres across on the ground, under the cover of a shrub.

But the typical shelter for the overwhelming proportion of rodents is not up a tree, on the ground, or indeed a pile of wood in the middle of a pond, but an underground burrow. However, are these burrows examples of animal building and, even if they are, does their completed design or the excavating of them show any degree of intricacy? The burrow system of the deermouse (*Peromyscus maniculatus*) is a single chamber at the end of a short tunnel. Not very exciting but it does at least have two parts to it: the entrance tunnel and the nest chamber. The burrow system of the woodmouse (*Apodemus*

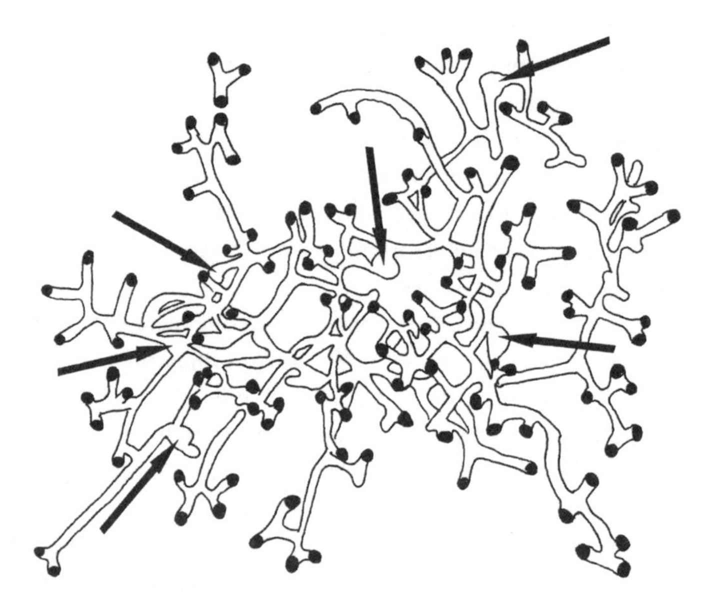

Figure 1.1. Brants' whistling rat burrow system: 115 burrow entrances (black dots) allow a foraging Brants' whistling rat to make a rapid escape from predators; arrows mark six nest chambers.

After Jackson, T. P. (2000). Adaptation to living in an open arid environment: lessons from the burrow structure of the two southern African whistling rats *Parotomys brantsii* and *P. littledalei*, *Journal of Arid Environments* 46, 345–55, Figure 1, part (a). © 2007 with permission from Elsevier

sylvaticus) usually takes the form of a loop of tunnel protected by the root system of a tree and containing possibly separate food and nest chambers. From this ring, there radiate maybe five or six burrows that exit on different sides of the tree, giving the mouse additional security when entering or leaving. This level of complexity alone forces us to take burrow digging seriously within the pages of this book, but it looks modest in comparison with the burrow architecture of Brants' whistling rat.

In one location in southern Africa where this rat species was studied, the *average* number of entrances to a burrow system was 41, and they are known to have up to 500 (Figure 1.1). You might guess that this rat is a highly social species, its burrows teeming with scurrying tails and the patter of tiny feet, but you would be wrong. Each system is normally occupied by a single rat, and the number of occupants rarely exceeds three. What is going on? Brants' whistling rat is an inhabitant of desert with scattered, stunted plants and very little cover from predators. The rat is a herbivore, feeding only on the sparse green vegetation, making it very vulnerable to predators when foraging. However, with a branching network of tunnels extending under the desert, which incidentally is probably not expensive in terms of energy to dig because of the soft sandy soil, and with multiple entrances, a rat is only ever a short dash away from safety.[3]

Naked mole-rats (*Heterocephalus glaber*) (which are indeed rats, not moles) are also inhabitants of the desert regions of southern Africa. Their branching burrow system can comprise more than a kilometre of tunnels often in hard compacted soils, but with few exits to the surface, and these rarely used. In this species such a system can easily be occupied by fifty individuals living in a complex society. Their burrow system is essentially their whole world as they rarely venture above ground. In this unvarying environment which is neither cold nor hot, the mole-rats manage without fur. This effectively makes them slimmer, allowing them to dig narrower burrows which therefore require the removal of a smaller volume of spoil and are consequently energetically cheaper to dig. These burrows are dug in search of food which takes the form of giant turnip-like tubers which, once located, can sustain the colony for several months.

It seems that we should certainly take burrowing behaviour seriously, in which case the rodents do represent the most important building abilities of mammals, at least in terms of their numbers. It is, in fact, hard to find much other building of significance occurring among mammals. Dogs, cats and mongooses as builders, antelopes, pigs and horses as builders? Not much more here than a bit of nest building and some modest burrowing. Over 20 per cent of all mammals are bats (more than 900 species), yet bat architects do not spring readily to mind, although perhaps surprisingly some of them are accomplished if unsophisticated builders. Typically these bat architects are inhabitants of wet tropical forests where they make shelters by biting through the woody veins of large palm leaves to create a structure a bit like an umbrella with all its spokes broken and hanging down. This leaves one major group of mammals yet to consider, the primates, that is the monkeys, apes and us. Of them I am tempted to say, like the hypothetical Martian, that apart from ourselves, there is little to impress. However, I need to be rather more careful.

Nest building in the primates, excluding ourselves, is confined to the great apes (orang-utan, gorilla, bonobo and chimpanzee). Chimpanzees routinely make night nests, which are generally used only once then abandoned. We would not ourselves put a great deal of effort into making such a temporary structure, and chimpanzees don't appear to either. Typically, the chimpanzee stands at a major fork in the branches of a tree and bends two or three branches in towards itself and then stands on them, fracturing them to create a platform which stabilizes as the branches splinter where they bend. Side branches on these broken limbs are now bent towards the middle, building up the platform, and a few additional leafy stems may be broken off elsewhere and added to the centre of the bed. The time taken to do this is generally less than five minutes. This might appear to show that nest building in chimpanzees requires little skill, but then I could argue the reverse: to build a nest so economically is evidence of skill. This is an issue that I will return to later in the book.

But, unimpressed as the Martian was by our ancestral primates, it was not their nest building which struck him as worthy of comment, it was the waving of sticks and the shaping of stones, that is, their tool use and tool-making. In fact a diverse range of species use tools, although their numbers are small, and a proportion of these even make tools. Asian elephants will use branches held in the trunk as fly switches. In a study of a small group of captive Asian elephants supplied with branches that were too big to be used as switches, more than half reduced the branch to a handy size, mostly by standing on the main stem and pulling bits off with the trunk. The elephants then used the trimmed branches as fly switches. Observations of this kind are biologically interesting because they invite us to believe that the tool user has some special insight into the consequences of its actions, even that it invented the idea. As you read this, you may find yourself charmed by the image of an elephant finally employing its skilfully crafted and carefully planned switch to take its revenge on the pestering flies. If so, let me ask you now as I shall do again in Chapter 7, what evidence would you put forward to justify that view of the elephant's mind?

But are tools a proper subject for consideration in this book? Tool-making involves construction behaviour—how could it not? So we certainly should assess not simply the importance of tools in human evolution but also whether the making and using of them by animals generally is something special. To do that it will be necessary to be ruthlessly objective. That is an approach I will try in Chapter 7; until that time, I am going to suspend judgement on toolmaking. That being the case, so far it would seem that the most consistently impressive group of mammalian builders is actually that of the rodents.

The number of bird species, at nearly 10,000, is twice the number for mammals, and the overwhelming majority of them build some sort of nest. These are located in all sorts of places: holes in trees or the ground, on cliffs and supported by or hanging from branches. The hanging nests are perhaps the most impressive. They need to have special attachments to secure the nest to an overhead support and be strong enough that the bottom does not split open, releasing the contents. The nest of the little spiderhunter (*Arachnothera longirostra*) hangs from the underside of a large leaf of some bushy plant found in Asian tropical rainforests. Looking at the upper surface of the leaf as you walk past, all you see are dozens of small white spots scattered

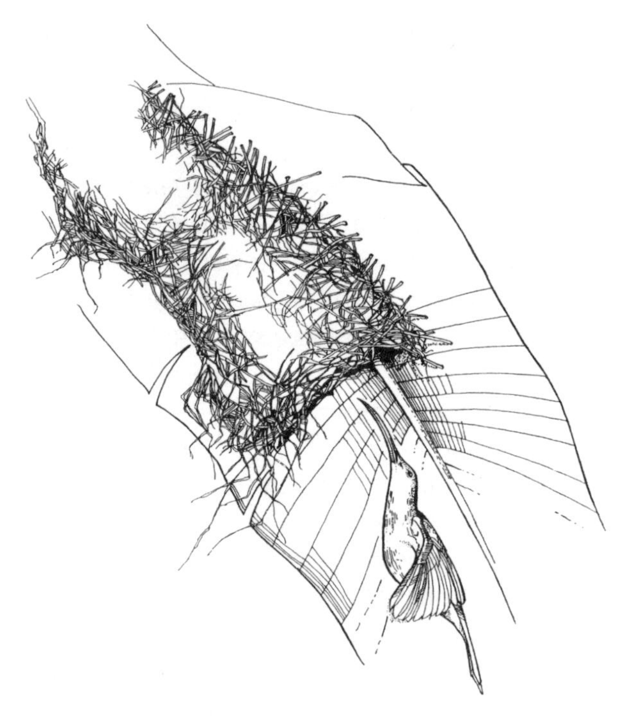

Figure 1.2. The nest of the long-billed spiderhunter, made largely of strips of plant material, is suspended from about 150 silken pop-rivets which the bird has driven through to the upper leaf surface with its beak.

M. Hansell, *Animal Architecture and Building Behaviour* (London: Longman, 1984)

over its surface; these are actually pop-rivets made of silk. The bird, which has a long, curved beak, can apparently hang under the leaf, and drive through it a tiny lump of spider or caterpillar silk with just the right force to create a hole through which the silk nodule simultaneously passes. As the beak is pulled back, the aperture in the leaf contracts. The lump of silk is trapped on the upper leaf surface while the bird still holds a thread of silk hanging down through the hole. This silk strand will become one of about 150 from which a nest will hang like an airship gondola (Figure 1.2). It is hard to say quite how difficult all this is as the nest-building behaviour of this shy forest-dweller has never been observed. My examination of the nest and leaf to which it is attached shows that the holes in the leaf are certainly struck from below, but my account is an interpretation. This spiderhunter nest appears to be one of the more complicated that birds make. Intricate nest structures of various designs abound among the birds, unlike mammals where the dam building of beavers is unique. There are also bird tool-users, even tool-makers, but more of them later.

In the remaining vertebrate groups, there is not a lot to report in terms of construction behaviour, let alone tool manufacture, by reptiles and amphibians (i.e. frogs, salamanders and company). The fish do show some interesting building and burrowing behaviour, and will feature again in later pages. To sum up the building activities of vertebrates then, for strength in number of species and depth in technique, the birds are the tops. It should not come as a surprise if important conclusions on the nature of animal building come from the study of them. Even so, the majority of species we will be looking at are not vertebrate but invertebrate builders. They have a huge advantage over the vertebrates in the number of species available for entry into any 'best builders' competition, but the best of what they can build rivals even some of our own achievements.

Several of the invertebrate phyla have very few species, less than a hundred, or, in some phyla, a single strange species may be given a whole phylum to itself. The larger the phylum, obviously the more likely it is to be mentioned here, but as the book progresses it will become more and more apparent that, as a source of examples, the arthropods totally dominate. This is in large measure because they constitute an extraordinary 75 per cent or so of all described animal species, vertebrate and invertebrate. Arthropods are everywhere—land, sea or air. They are creatures characterized by external skeletons and jointed limbs, and familiar to us as house flies, spiders, woodlice, crabs, scorpions and the like. It is a formula that has proved impressively versatile and adaptable.

The majority of arthropod phyla are found in a dozen or so classes with technical names that are unfamiliar to most, including professional biologists, but, sticking to common names, the most notable arthropod builders are found among the spiders, where nearly all species build something, and the insects, where building is widespread. Of special interest are the so-called *social insects* because they often live in large groups and build nests that can house the whole colony. A worldwide example of this is the domesticated honeybee.

A honeybee nest may contain around 10,000 adult insects that have constructed hexagonal cells made of wax for the storage of honey and for the rearing of their maggot-like larval stages. This is an extraordinary development of social living that has no equal among vertebrate animals other than ourselves. In some aspects of their building, social insects even surpass us, for example in the scale of their structures. The nests of South American leafcutter ants (*Atta vollenweideri*) are subterranean labyrinths extending as much as 6m under ground and containing as many as 8 million adults (perhaps another 2 or 3 million for larval stages and eggs) (Figure 1.3). This one colony therefore equates to the population of the largest human cities, living all in the one structure. The colony is sustained by the constant import of freshly cut pieces of grass leaf which are not eaten directly by the ants but chewed to a pulp and used as compost to nurture fungi grown in special fungus gardens to provide the colony's food. Exhausted garden waste and other colony debris is dumped in huge underground silos.[4]

This vast structure of underground passages and chambers, domestic and horticultural areas needs to be ventilated, and a system to accomplish this is built into the structure as well. At the surface, the excavated soil forms a shallow mound on which are numerous entrance holes, topped by small earthen turrets. The passage of wind over the mound surface is sufficient to cause a lowering of air pressure over its highest part compared with the lower parts. This causes air

Figure 1.3. Leafcutter ants' nest: excavation of the chambers and highways made by the leafcutter ant (*Atta laevigata*) in Argentina after the nest had been flooded with 6.7 metric tonnes of cement mixed in 9,000 litres of water.

Martin Bollazzi

to flow out of passages that exit at the top of the mound and into passages at the edge. This *induced flow* mechanism of nest ventilation exploits the same principle that causes lift on aircraft wings but, in the case of the mound, a pressure difference is created regardless of wind direction. Incidentally, although of all the architectural prowess shown by leafcutter ants we may be most impressed by the nest ventilation mechanism, the same principle is also shown by the mounds of some termites, the burrows of some rodents and, in flowing water systems, by some burrow-dwelling fish and mud shrimps. All have multiple burrow exits at least one of which is raised above the level of the others by the creation of an artificial mound. Are all these species cleverer than their relatives or is the evolving of a ventilation system relatively easy? More on that in Chapters 3 and 4.

We can now see where in the animal kingdom builders are to be found, but it is also worthwhile considering very briefly what these animals build for. Overwhelmingly, animals build homes. I mean 'home' in a rather general sense as a secure refuge, protected to some degree from the physical hazards of extreme cold and heat and the biological hazards of predators. Nests, burrows, cocoons, all are homes in this sense. They may of course be more than simply secure places. Like our own homes they may have additional features: food stores, waste disposal and even food production areas.

There are essentially only two other functions of animal-built structures: as traps or as displays. In terms of the numbers which build them, these two are far less important than homes, but both raise challenging issues, so I want to discuss them in a lot more detail. What creatures build these structures? Well, the most obvious trap builders are the spiders. We think of the orb webs of late summer, but across the spider species there are a lot more trap designs than that, and there are certainly trap builders other than spiders. Display structures may be hard to think of at all but you have very likely heard of the 'bowers' built by male bowerbirds. These are not nests and male bowerbirds have nothing to do with nesting. The structures they build are simply to attract females. Bowerbirds are virtually the only animals to make such display objects and their remarkable elaborateness may tell us something about ourselves, but that needs to wait till the last chapter of the book. The job of the next chapter is to look at the consequences of builders altering the world in which they live.

2

Builders Change the World

'Wombats Detected from Space' is the title of a scientific paper published in 1980 in the journal *Remote Sensing of Environment*.[1] The burrowing and associated soil disturbance, it goes on to explain, damages the surrounding vegetation to such an extent that, even with the satellite imaging available a quarter of a century ago, it could be detected and mapped from space. The southern hairy-nosed wombat (*Lasiorhinus latifrons*) is a sturdy beast about 1m (3.3ft) long, weighing up to 30kg (66lb). For burrowing in the compacted desert soils of Southern Australia, it also has powerful legs and strong claws. A single wombat burrow is not a very complex or extensive thing, 6 to 8m of tunnel with generally one, but sometimes two, openings. Burrows, however, usually occur in clusters or warrens amounting to over 80m (263ft) of tunnel and twenty or more entrances. This produces a distinct mound, rising a metre or so above the flat surroundings. An individual warren may be 30 or so metres across and warrens themselves may be clumped. The result of this disturbance is bare patches in the scrub landscape of a few hundred square metres or as much as one square kilometre, easily visible from space. This chapter examines how, in so many ways that we fail to notice, builders change the world.

Incidentally it is worthwhile noting what the wombats use their burrows for, as different species use burrows in different ways. Of course, a burrow provides a secure environment buffered against extremes of climate and a refuge against predators, but the wombat is an extreme animal, extremely idle; this makes a secure home particularly important in its life. Hairy-nosed wombats could be described as nocturnal but to say so is a bit misleading. An average day for a southern hairy-nosed wombat is: spend over twenty-one hours in the burrow, emerge during darkness for no more than two-and-a-half hours to feed, cover an average of 200m (a distance covered in twenty seconds by Olympic athletes), before going to ground again. It is not an energetic life and it is energy or rather the lack of it that is a wombat's problem. Its diet consists of grasses and sedges, which for most of the year are tough to digest and low in nutritional value. The wombat solution is to move little and slowly, and use the long periods in the burrow for patient digestion.

What about the cost of burrowing? Given the hard soil, this could be high, but that depends upon whether wombats actually do much digging of fresh burrows rather than mostly low-cost maintenance in a warren that has been on that site for, say, 100 years. Is a warren likely to be that age? Well, all right, let's say 1,000 years then. I don't actually know the possible age of a warren, but neither does anyone else. However, it has been seriously suggested that some burrow systems of the European badger (*Meles meles*), referred to as setts, may be several hundred years old.[2]

Badgers in Britain, surprisingly for a largish carnivore, feed mainly on earthworms. Living in quite productive habitats for earthworms, badgers have little need to change the location of their sett. Over time, a sett can be gradually extended to become very large. One partially excavated in England a few years ago had 879m of elaborately branching burrow, 50 chambers and 178 exit holes. This will have probably been occupied at any one time by a mere handful of badgers. The question here is whether this same sett had badgers in residence when George III lost the American colonies in July of 1776, or even on 14 October 1066 when William, Duke of Normandy, beat King Harold at the battle of Hastings, making way for the Norman conquest of England.

In Old English, the name given to badgers was 'brock', a name now given as a surname for badgers in children's stories. If you look at a basic road atlas of mainland Britain you will probably find twenty-five or more place names that begin with either 'badger' or 'brock', Badgers Mount in Kent for example, and Brockholes in West Yorkshire. It is completely unscientific to say so, but I like to imagine that the permanence of a badger sett in the landscape led naturally to the naming of farms, hamlets and then villages to indicate their proximity to a major badger sett. This is not to say that the tunnel system now present in a particular sett is the one present hundreds of years ago. Rather, it indicates that the site has been permanently occupied by badgers for that time, that parts of the system are traceable to its ancient layout, and that the landscape we now see round about it owes something to its long association with badgers.

Let us think of a human equivalent, for example the church of St Peter and St Paul in the village of Brockdish, just on the Norfolk side of the gentle River Waveney, which marks its boundary with Suffolk. It is a church largely rebuilt in the 1870s, as can be seen from the splendid Victorian tiles that decorate the wall behind the altar. However, parts of the mediaeval church remain: a piscina (for washing up vessels used in the Mass) dating from the thirteenth century and one window of Saxon date, a remnant from the tenth century. This church, although continually changing itself, has been exerting an ecological influence on the local landscape for a thousand years, and the shape of the village and surrounding fields bear witness to this. If any local badger sett were of similar age, we should expect that it also, in its smaller way, would have engraved some mark of its history on the local ecology. There is in fact evidence from some English badger setts that tonnes of soil have been excavated over time, and that there is a typical vegetation associated with these areas of disturbed soil: one of elder bushes (*Sambucus nigra*) and stinging nettles (*Urtica dioica*). Builders, human and badger, do change the world.

Looking round at our landscape now, we see so much of its ecology dominated by human activity, but we have only very recently appeared on the planet and our history as major ecosystem engineers only extends over a few tens of thousands of years. Other creatures have been altering landscapes for tens, even hundreds of millions of years before us. Can we see physical evidence of that? If so, how great and from how long ago? Some animal built structures are very fragile, but others are durable. A burrow, for example, can become fossilized when filled with a deposit of a different type from the soil in which the burrow was dug. Such 'fossils of behaviour' are referred to as *trace fossils* and they are very common in some rocks, for example of former seabed sediments.

A sandstone deposit recently excavated in South Africa was found to contain the fossil remains of a system of branching tunnels dug by a land-dwelling vertebrate.[3] These were so well preserved that they still showed the scratch marks of the animals that dug them about 240 million years ago. The widest of the tunnels, which were about 15cm (6in) across, had two grooves in the floor, one each side of a central ridge. This suggests that these tunnels had two lines of traffic and therefore that the burrowers lived socially. This is supported by the discovery of fossil skeletons in one of these burrows of around twenty individuals of a small mammal-like reptile of the genus *Trirachodon*. The jumble of skeletons and the sediment filling the burrow suggests that they were all drowned together when their home was overwhelmed by a sudden flood.

This is an interesting snapshot of an ancient tragedy, but have these burrows and their subsequent traces significantly altered the local ecology? Possibly not very much, but consider this, on the face of it, modest environmental modification: some small lumps of mud found on the walls of rock overhangs and caves in the Kimberley region of Western Australia. These are the remains of the nests of mud dauber wasps (probably *Sceliphron* species), hardly worth a glance were it not for their association with prehistoric rock paintings. This led to a dating of the wasp nests by subjecting the quartz grains of the mud to a technique called Optically Stimulated Luminescence (OSL). This gave an age for the mud nests of 17,000 years.[4] That is certainly an exceptional feat of endurance for something so apparently insubstantial as a small wasp nest, but the question we are concerned with here is whether these blobs of mud really alter the local ecology. I'm not sure, but I know some blobs of mud that do. Nests of mud dauber wasps located on walls under bridges in the United States are, because of their firm attachment to the vertical concrete surfaces, used by barn swallows (*Hirundo rustica*) and eastern phoebes (*Sayornis phoebe*) as foundations on which to build up their own mud nests. If that sounds to you just a modern effect, then I give you the 2kg (4.4lb) mud nest of the white-necked rockfowl (*Picathartes gymnocephalus*) in Ghana. This, the heaviest of all rock wall-attached mud nests is, by preference, built where the remains of mud dauber wasp nests help secure it to the difficult surface.

These examples, the badger setts and the mud dauber nests, show the surprising endurance of built structures on the environment, but are not the best examples to show how builders can actually change the appearance of landscapes. In parts of Washington State and California in the USA there is a grassland landscape characterized by regularly spaced shallow mounds about 0.8m in height and around 15m across. Once thought to be the product of physical forces such as repeated freezing and cooling, it is now clear that they have been created by burrowing rodents such as the pocket gopher (*Geomys*). Above ground, the gophers defend territories around their burrows; below ground they maintain separation of their tunnels by listening to the vibrations of their burrowing neighbours. The result is an even spacing of gophers. This, together with the excavation of spoil from the burrows, leads over time to the even spacing of mounds across the landscape.

How long it takes for this landscape to be created we have no clear idea, but it also occurs in the plains and desert scrub of Argentina where it is created by the burrowing activities of rodents such as the tuco-tuco (*Ctenomys*). This type of animal-generated landscape has been given its own special name, *mima prairie*. In South Africa there is a similar phenomenon—regularly spaced mounds in this case occupied by both common mole-rats (*Cryptomys hottentotus*) and by the termite *Hodotermes viator*. These mounds, which average 2m height and about 28m across, apparently come about because in both the rodent and the termite colonies, there is competition between neighbours, while the termites and rodents within each mound tolerate each other. One study has carbon dated material from the core of a mound. This gave a date of between 4,000 to 5,500 years ago for their foundation. Each mound is a living site built on its own archaeology, like the hearts of many human cities.

The effect of even small organisms altering landscapes on this sort of time-scale can be impressive. In Botswana, in southern Africa, there is a landscape of regular corrugations with parallel gullies about 50m apart separated by ridges about 2m in height. Not especially impressive at ground level but, from the air, ridges can be seen to extend for up to a kilometre. These landscape features are of a different order of magnitude from mima prairie mounds, but current interpretation is that they are created by termites of the genus *Odontotermes*.

This phenomenon of habitat modification has been termed ecosystem engineering and the creatures responsible for it, ecosystem engineers. There are some problems with the use of this term, which it is as well to be aware of. Any living thing, critics of the concept argue, alters the landscape to some degree; a tree creates shade, alters soil moisture and nutrient availability, but from the point of view of this book we can ignore that. Here we are concerned with a specific aspect of ecosystem engineering; how animals *through their building behaviour* physically alter the environment. These effects alone are significant enough to make ecosystem engineers worth studying.

We should again acknowledge a debt to Charles Darwin. No, it is not another tribute to his insights on evolution. The acknowledgement is for a book Darwin published in 1881, shortly before his death, *The Formation of Vegetable Mould through the Action of Worms with Observations on Their Habits*. Now, we can recognize this as an early study on the ecosystem engineering of that burrowing creature, the earthworm. At Down House in Kent, where Darwin lived in secluded domesticity for forty years, he conducted observations and experiments on all kinds of biological phenomena, orchid pollination, carnivory in plants and, particularly in his later years, on the burrowing of earthworms.

Earthworms feed by passing soil through their guts, which may then be excreted at the soil surface like the squeeze of a toothpaste tube—a worm cast—much to the distress of the keepers of golf greens. Darwin collected and weighed these bits of earthworm excretion over a period of time on a particular patch of ground, did the arithmetic and came up with the figure of 8.4 pounds per square yard per year (that is an impressive 4.6kg per sq m per year). He estimated that this amount of soil movement could, over time, effectively bury archaeological remains. This was no idle armchair exercise. He conducted an experiment with the assistance of his son Horace, to see how quickly a large stone would disappear into the ground as it was at once undermined by earthworms and covered by their casts. On one occasion, attending the excavation of a Roman villa concealed beneath Surrey farmland, he noted that as the archaeologists exposed the structure so also did earthworms in their small way cover it again with their casts. From the number of casts detected during the next seven weeks over the area of the atrium floor, he was able to estimate the worm population under it and, from Roman coins found during the excavation, calculate that in the 1,500 years since it had been abandoned, the depth to which the villa was buried could be simply the consequence of earthworm activity. That is a remarkable and pioneering example of field ecology. It is also another example of the extent to which builders can, surreptitiously in this case, change the world. However, the evidence presented so far this chapter, of how builders change the world over time and space, fails to make a key point: the extent to which the environments created by builders alter the world not only for themselves but for a mass of other species. It is that important effect which we now need to explore.

You will yourself have noticed, as you walk across a beach exposed by the outgoing tide, similar worm casts to those observed by Darwin. Many species of worm, as well as other kinds of creatures, burrow into the muddy sediments of the sea floor. The population densities of these burrowing worms can be very high. A figure of 5,000 per sq m has been calculated for the predatory polychaete worm *Nereis (Hediste) diversicola* and 50 per sq m for its near relative the lugworm *Arenicola marina*, which feeds in a similar way to the earthworm. These marine burrowers are therefore, like earthworms, moving the sediment, although in ways that differ between species. The mud shrimp *Upogebia stellata*, which filters the water current to obtain fine food particles, digs a burrow down to about 30cm, bringing the spoil up to the surface; another mud shrimp *Callianassa subterranea* digs right down to 90cm. Other burrowers move the sediment in the reverse direction. The worm *Maxmuelleria lankesteri*, for example, feeds by leaning out of its burrow to skim off the freshly deposited surface layer of sediment within its reach which, after passage through its gut, is deposited at a depth of about 80cm. A study echoing that of Darwin calculated from the natural density of mud shrimp (*Callianassa*) burrows, and the rate of burrowing in an aquarium, that they must be bringing to the surface of the seabed 15.5kg dry weight of mud per sq m per year. The marine mud, apparently passive to a passing Scuba diver, is in its top metre at least, a restless, dynamic ecosystem.

The very high densities of creatures living in marine mud are an indication of what a rich food source it is. This is something of a puzzle since mud is notably lacking in oxygen, upon which organic decomposition and hence nutrient release depends. Lack of oxygen in waterlogged deposits is after all why hundreds of human bodies, some dating back 10,000 years, have been recovered from European peat bogs, one (the 2,000-year-old Tollund man from Denmark) was so well preserved that the local police were called in to investigate his murder. It has been shown that in undisturbed marine sediments the depth to which the oxygen-loving bacteria responsible for most organic decay can flourish is a mere 1 to 6mm. But the mud is disturbed; it is being penetrated and redistributed by burrowing crustacea, worms and bivalve molluscs (clams). By their activity the mud environment is transformed.

Lugworms ventilate their burrows actively, by driving water through with their body contractions, bringing dissolved oxygen not only for their own gills but incidentally to oxygen-loving bacteria. The mud shrimp *Callianassa truncata*, a creature of only 2cm in length, lives in a complex arrangement of chambers and tunnels that penetrate to a depth of 50cm. All that can be seen of *Callianassa* at the mud surface is a view of one or two funnel-shaped depressions beside a mound of mud with a hole in it (Figure 2.1). This is the marine equivalent of the prairie dog burrow ventilation system. As fluid, in this case water, passes over two apertures connected by a tunnel, one placed higher than the other, a pressure difference is created such that the water is drawn out through the top of the mound and into the burrow through the funnel. This burrow system is passively ventilated by the induced flow, bringing oxygenated water deep into the mud.[5] Marine mud is rich in organic material, but it is the habitat modifying powers of the burrowers which bring about the release of energy and nutrients. This tells us something very important about the density of lugworms in the mud sediments. It is due in a substantial way to the activities of the lugworms themselves. They ventilate the sediment with oxygenated water, allowing a population of oxygen-loving bacteria to break down organic debris, releasing nutrients. The bacterial population flourishes and, as a consequence, supports a food chain of protzoans, diatoms and nematode worms, enriching the mud on which the lugworms feed.

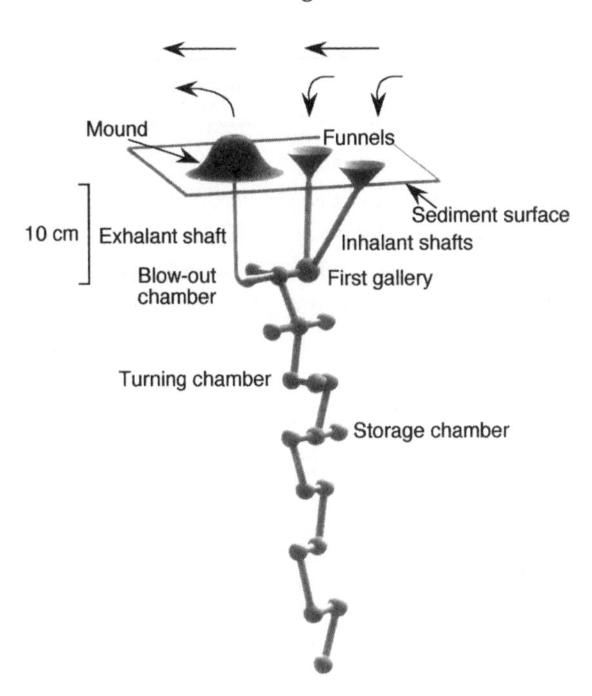

Figure 2.1. Mud shrimp burrow: water is induced to flow into the mud shrimp burrow by the 'mound' and 'funnel' openings; this ventilates the tunnels, which penetrate deep into the sediment.

Adapted by permission from Macmillan Publishers Ltd: Nature, W. Ziebis, S. Forster, M. Huettel, B. B. Jørgensen, Complex burrows of the mud shrimp *Callianassa truncata* and their geochemical impact in the sea bed. Nature 382 Aug. 15, 619–22 Figure 2 1996 © 2007

These examples show that burrowing creatures, by introducing a new complexity (their burrows) into the mud, can increase its productivity. But surely they must also be introducing an additional level of structural complexity into an environment—microhabitats (niches) for other kinds of organisms. Is there evidence that burrowers, or indeed any animal builder, can enhance biodiversity through ecosystem engineering?

The sand tilefish (*Malacanthus plumieri*) is found off the Caribbean coast of Colombia in the rather featureless sandy areas that occur inside and beyond the coral reefs. In contrast to the obvious richness and diversity of species on the reef itself, marine life in this habitat seems very limited. However, it is significantly boosted by this one species of fish.[6] Each male tilefish generally shares a burrow with a female harem and, since males are territorial, burrows are fairly evenly spaced. This is not, however, a story of the effects of the burrows themselves on biodiversity, because over each burrow the sand tilefish also build a mound of several thousand pieces of coral rubble, stones and mollusc shells that is up to 1.5m in diameter and 25cm high.

What purpose the mound serves for the tilefish is not clear (protection of the burrow during storms, sexual advertisement?), however, a study of these mounds has found them to be occupied by thirty-two other species of fish of diverse ecology: herbivores, debris feeders and carnivores. Some of these fish only occur in the mounds as juveniles, showing that the rubble mounds act as nurseries for species that as adults live in other habitats. The species diversity in the mounds is of course not confined to fish; many species of invertebrates were also found to be present: marine worms and snails, sea urchins and brittle stars, various crab species and other crustacea. Here is an example showing not only that animal-built structures add variety to the habitat, but also that the builders in creating new niches induce other species to join them. This in turn further adds complexity that may draw in additional species, predatory fish for example, to feed on one of the other species of immigrants.

Let me not get too carried away. Strictly, I am making an assertion based on the correlation between habitat complexity and species diversity. What as a scientist I would like to do is present you with the results of an experiment in which diversity is compared between an experimentally manipulated habitat and an unaltered control. Here is such an experiment, elegant yet so simple in equipment terms that you can repeat it yourself if you wish. All you need is a few paperclips. This experiment concerns the effect on local biodiversity of folding over or rolling up a leaf. There are a number of species of insects and some spiders that do this to make shelters. One such is the caterpillar of the small moth *Acrobasia betulella*, the 'birch tube-maker', which rolls up a birch leaf to make a retreat to use when it is not feeding on the neighbouring leaves. Examination of these rolls shows that they may be occupied by more than one caterpillar not always at the same growth stage, suggesting that opportunists may make use of shelters already made by another. The rolls may also be occupied by caterpillars of other species, evidence suggesting that leaf rolls enhance local biodiversity. But let's test that with an experiment.

Such an experiment compared diversity on branches of cotton wood (*Populus* species) where some leaves were rolled up and fastened with a paper clip, with unaltered branches acting as controls.[7] Creation of these simple leaf rolls not only resulted in seven times the abundance of insects on 'experimental' branches compared with 'controls', but also a rather staggering four times the species diversity. Many of the colonists exploiting the leaf rolls were not leaf feeders themselves but predators. A supplementary experiment, which attached paper rolls in the experimental patches instead of rolling the leaves, showed that even for leaf feeders, the paper rolls were attractive as shelters.

The examples of the tilefish mounds and caterpillar leaf rolls reveal species diversity prospering where islands of refuge are spread through habitats of relative exposure. This may well be one of the effects of gopher burrows in mima prairie, but closer examination shows a variety of less obvious ways in which the patchiness they produce in an otherwise rather uniform environment creates opportunities for species diversity.

The activity of the resident rodents in mima prairie is focused on the mound where an individual comes and goes and deposits freshly dug soil. This constant disturbance permits little vegetation to grow. However, beyond this central bare patch there are subtly different concentric zones of vegetation. The first is a zone of more stable soil, enriched by nutrients released from recently dug soil and the excreta of the mound occupant, resulting in luxuriant plant growth. Beyond this, the plants show rather weak growth due to lower levels of nutrients and the shade cast by the vigorous growth beside them. Beyond this zone, with increased access to light, plant vigour partially recovers. These zones of course are not just zones of plant luxuriance but of plant differences, different species preferring different conditions. This in turn results in different communities of insect herbivores, and insect and spider predators. There is in fact an additional influence on diversity because the patch system in mima prairie is to an extent dynamic. If a mound owner dies as the local population is in a phase of decline, he or she may not immediately be replaced. This is an opportunity for pioneer plant species to colonize the bare patch at the centre of the mound. You might suppose that, because such patches are small, one specialized plant species would be able to out-compete all rivals for these sites. However, this is not likely to be the case. Instead, simply because the patches are small, the first species to start growing on the patch then excludes the latecomers. By chance, the seeds of different species arrive first at different bare patches, promoting diversity.

The evidence is strong that builders enhance biodiversity by increasing habitat complexity, but is it not possible that animal builders will destroy habitats so reducing species diversity? This is most likely to be the case where, unlike the grassland plains or seabed sediments that we have been concentrating on, the habitat already has high diversity. One such is the forest and stream habitat of the beaver, and there is indeed some evidence that beavers do reduce biodiversity. They feed on deciduous or broadleaved trees, to the extent that they may reduce them at the expense of conifers. By blocking rivers and dams beavers may also destroy the spawning areas of some fish or disrupt the migration paths of others. But, on the other hand, they also bring about changes that promote diversity. Their felling of trees can create clearings where flowering plants can flourish, encouraging insects that attract additional species of birds. The still water behind the dam is habitat for planktonic crustacea and mosquito larvae. Planktonic feeding duck species, such as teal, benefit as a result. In winter, adult mosquitoes can shelter in the beavers' lodge alongside the residents, biting them from time to time for a blood meal. The net effect of beaver ecosystem engineering is still probably to enhance biodiversity.

So, builders attract and benefit other species, but to what extent do those other species depend upon the builders to provide the environment they need? This question clearly has implications for conservation. If many species are utterly dependent upon builders for their homes, the loss of the builders from a habitat could lead to a dramatic decline in biodiversity. What, also, is the relationship between the builder and the squatter? Does the builder ever benefit, or is the relationship merely neutral? Do some squatters exploit their hosts, or even cause them harm?

To start answering those questions, we need first to consider the species diversity associated with the tubes built by a marine worm by the name of *Phyllochaetopterus socialis*. Worms of this species, as its second name suggests, clump together and, in so doing, form dense tangled masses of dwelling tubes in an otherwise rather uniform muddy substrate. A study, which looked at all organisms greater than one millimetre in length associated with these tube aggregations, came up with a list of sixty-eight species, mostly crustacea and molluscs. However, the majority of these could also be found in other habitats in the area. Their association with the worms was non-specialized and opportunistic. Builders, as we have established, are mostly makers of shelters; leaf rollers, pocket gophers and sand tilefish all exemplify this. A shelter in its simplest form is just a barrier between the organism and a hostile world beyond. It is generally a quite unspecialized microhabitat and one that many non-builders can therefore take advantage of, although they may also find alternatives elsewhere.

Some species, however, are more dependent on shelters made by specific builders. In the steppe plains of China and Tibet, two species of snowfinch (*Montifringilla*) are largely dependent for nest sites on the burrows dug by a relative of the rabbit, the pika (*Ochotona*). A better-known example of a bird that nests in mammal burrows is probably the so-called burrowing owl (*Speotyto cunicularia*). This, in spite of its name, is not much of a burrower but depends upon mammal burrows for nest sites. In some grasslands of the United States, for example Oklahoma, it depends heavily upon the extensive burrow systems of the black-tailed prairie dog (*Cynomys ludovicianus*). Black-tailed prairie dogs are found from Montana in the United States down into Mexico, but the burrowing owl also occurs in grassland and savannah habitats in South America. In Argentina it can be found nesting in association with that moustachioed and bewiskered rodent, the plains viscacha (*Lagostomus maximus*), the maker of complex burrow systems which extend over several hundred square metres and have anything up to forty entrances. These burrow systems are known locally as *viscacheras*, and some, incidentally, are thought to be several hundred years old.

Other bird species also make use of mammal burrows for nest sites, but some do in fact dig their own homes within those burrows. In grassland areas of Nigeria and Tanzania the sooty chat (*Myrmecocichla nigra*) can be found nesting inside the burrows of aardvarks, a substantial mammal of 70 kilos or more, which has powerful claws not simply for digging a burrow, but also to demolish the mounds of the termites on which it feeds. The chat, weighing no more than 40g, avoids accidental damage of its nest from the aardvark by excavating a nest cavity in the roof of the burrow. Exactly the same solution has been arrived at in South America by the common miner (*Geositta cunicularia*) which digs a tunnel in which to make a nest. One of the sites where it chooses to make its burrow is inside a viscachera. These bird species may not be entirely dependent upon mammal burrows for nest sites but their ranges may well be extended by using these secure sites where few are available. It goes without saying that the nest cavities dug by the birds offer habitats for something else. The cavities dug by common miners in viscacheras are sometimes taken over by blue and white swallows (*Notiochelidon cyanoleuca*), a species we shall come across again as an opportunist cavity nester that utilizes a variety of sites. Some birds exploit less obviously useful sites of animal construction, spider's webs for example. This requires some explanation.

Spiders as a general rule lead solitary lives, predatory towards most species, cannibalistic when the opportunity arises. However, there is almost no rule in biology without there being exceptions, and there are a number of species of spider, largely in the tropics, that live in colonies of hundreds or thousands within complex spaces enclosed by sheets of silk draped between branches. These structures are large. The African species of social spider (*Agelana consociata*) weaves a mass of sheet webs that envelope a volume up to 3m across—large enough to envelope both you and I, let alone a bird nest. In Africa, three species of sunbird (small birds similar in appearance and habits to the hummingbirds of the New World, i.e. of The Americas) are recorded as nesting in such web complexes, as is also another species, the little grey alseonax (*Muscicapa epulata*). In South America, the royal flycatcher (*Onychorhynchus coronatus*) appears to make use of social spider's webs too. This may sound a little vague but this reflects the vagueness of the descriptions we so far have. More study of what lies within social spider's webs will I am sure increase our list of birds that nest there and of other squatter species besides. What do the birds gain from this? Well, I'm not aware of any bird or mammal species other than our own that is afraid of spiders, and I don't think the social spiders themselves pose much threat to a predator of bird chicks such as a snake or rat. However, these spider's web complexes certainly conceal the bird nests, and smaller vertebrate predators may well be put off by getting swathes of silk web wrapped around their faces.

Another kind of socially living invertebrate whose homes are used as nest sites by birds is the termite. Here we have a lot more solid evidence of the bird species involved, chiefly parrots and kingfishers. Around the tropical grassland areas of the world are mound-building termites, notably the *Macrotermes* species in Africa and the *Amitermes* species in Australia. Their mounds can easily reach the height of the tallest humans at around 2m, and may reach 6 or even 7m, significantly taller than that British unit of measurement, the (sadly missed) Routemaster double-decker bus, height only 4.4m. The material of which termite mounds are made is predominantly soil mixed with faecal cement to form a tough composite. In the tropical forests of Asia, Africa and the New World there are also smaller, globular termite nests located in trees. All these are potential nest sites for a variety of bird species that peck through the outer wall of the nest and carve a cavity through the softer material that divides the interior into tiny chambers. At least seventy species of birds are known to do this, parrots by nibbling, kingfishers by chiselling.

Looking at the consequences to builders of their companions, the nest of the sooty chat seems unlikely to bring any significant disadvantage to an aardvark by excavating a nest cavity in the roof of its tunnel, nor will the common miner inconvenience the viscacha. Some relationships between builder and joiner even seem to be mutually beneficial. At least twenty-nine species of fish of the goby family are found in association with thirteen known species of burrow digging, snapping-shrimp (*Alpheus* species), one fish and one shrimp per burrow. The fish gets the shelter of the burrow, contributing little to the digging; it may also benefit by being groomed of ectoparasites by the shrimp. The grooming of course provides the shrimp with a few morsels of food, but more particularly the shrimp gains intelligence of approaching danger through contact with the fish using its long antennae, some goby species signalling alarm with specialized fin flicks.

A more compelling example of mutual benefit is not between builder and cohabiting animal, but between builder and associated fungus, the builders in this case again being termites. The *Macrotermes* species of the African savannah overcome the problem of digesting tough grasses with their own tiny guts by outsourcing the digestive responsibility to fungi. These fungi, which belong to the genus *Termitomyces*, have evolved a specialized relationship with the termites. The termites cultivate the fungi in special horticultural areas of their mounds, the fungus gardens. The fungi of course benefit in turn by being propagated by the termites. Indeed every potential queen that leaves the natal mound with the aim of founding a new colony, carries a gut rich in fungal spores, the innoculum for new fungus gardens.

Natural history is littered with examples of good solutions to problems that have evolved independently in different organisms. Even so it is very satisfying to find that this is also true of the unlikely habit of fungus cultivation by insects, which is found not only in termites but in the only distantly related ants. The ants in question are the leafcutter ants (*Atta* and *Acromyrmex*) of the New World, which cut fresh green leaves to compost in underground fungus gardens. As with the termites, the fungi themselves have evolved a specialized relationship with the insects, but they are different fungi. The fungi of leafcutter ant colonies are not closely related to *Termitomyces*, belonging to a different family, the Lepiotaceae.

The digestive problem the ants have is, to be precise, rather different from that facing the termites. In the forests of the temperate regions, such as much of Europe and North America, trees can get rid of their burden of leaf-chomping insects by shedding their leaves in the autumn. Next spring the insects have to get established all over again. In the wet tropics trees can remain green throughout the year, so they have evolved chemical countermeasures as one of their main methods of limiting damage from leafeaters. The lush green vegetation of tropical forests, for all that it looks inviting, is generally rich in toxins. The fungi break down the toxins and prosper; the ants eat the fungi and multiply. Such a mutualistic relationship between nest builder and occupant is not so much between the built structure and the joiner as directly between the two organisms. A large leafcutter ant nest can perhaps be thought of as a city-state from which the ants stream out into the surrounding forest or savannah to harvest green leaves to nourish their fungi. Their impact on the local ecology can be great; it has been estimated that in some tropical forests, *Atta* species alone will consume around 15 per cent of the plant growth.

Figure 2.2. Magnetic termite nests: the 'high-rise' profiles of magnetic termite mounds dominate this Australian grassland landscape, their flattened faces catching the evening sun.
Martin Harvey/NHPA

One place where it is easier for us to visualize directly the influence, even dominance, of large social insect colonies over ecosystems is in parts of the Cape York Peninsula of Queensland and the Northern Territories, in Australia. Here, in the dry season the mounds of the 'magnetic' termites (*Amitermes meridionalis*) stand, evenly spaced, their flattened profiles uniformly oriented like the tombstones in a giants' graveyard, casting their long shadows across the golden grass where, in the wet season, they stand in a shallow lagoon, their 3m high outlines reflected in the water (Figure 2.2). Their flat surfaces face East and West, or in other words their long axes are aligned North–South, hence 'magnetic'. This very specific mound orientation was shown, over thirty years ago, to be for temperature control. The flat Eastern face is warmed by the rising sun, the Western face by the setting sun; in the heat of midday, the upright slab of the mound that tapers to a sharp jagged edge at the top presents only a small area to the sun's rays. This ensures that nest temperature rises quickly to a useful 33 to 34°C, which it maintains with little variation until the evening. At night the termites radiate out from their mounds, to gather pieces of dried grass which are stored in chambers back in the mound, stores that can be used during the dry season and in the rainy season, when each mound becomes an artificial island.

I, and many others like me, have found this a convincing and convenient story of functional design to tell the students over the years, but it now seems that it may not be true, or at least not entirely true. The problem is that other *Amitermes* species in Australia and *Mactotermes* species in Africa build tall mounds, but all are rounded, lacking the dramatically flattened shape of magnetic termite mounds. The most obvious special feature of the magnetic termite habitat is its exposure to seasonal floods. Researchers are now investigating whether the flattened mound shape is a device to increase the surface area of the mound relative to its volume, to allow rapid drying of the stored hay after rain. So, perhaps the flattened mound design is not ideal for temperature control, because it results in rapid cooling within the mound after sunset, but is the best possible design given the constraint of keeping the food stores dry in the wet season.

Foraging trips into the countryside from the metropolis of a magnetic termite mound costs the residents time and energy. The further away that food has to be collected, the less net value it therefore has, and the more likely it is to be of greater net value to a neighbouring mound. Over decades, centuries perhaps, economics rather than warfare have spaced these cities across the land; some colonies die out, their towers crumble and fade back into the landscape. A young royal pair, queen and king, fortuitously land in unexploited terrain. They found a new colony in a new mound, the growing profile of which symbolizes their local control. Gradually, a quite even spacing of cities settles into the landscape. It is a world fashioned by termites. This reads a bit like the opening passage of some fantasy novel but it is nevertheless a fair description of the ecology of that Northern Territories and Queensland habitat. The energy of the sun, locked up in new plant growth after the rains is channelled through the termites to rematerialize as building work and new generations of termites. They, the buildings and the termites, in turn provide respectively places for other organisms to live and prey for other animals to hunt. The termites bring continuity and stability to the land. Within the mound, they have blunted the peaks and troughs of the daily temperature cycle and, over the course of the year, they have ensured continuous food availability in a habitat which shows only a brief phase of explosive productivity after the rains.

How long have habitats shaped by social insects been around? For ants we have some evidence from fossils. In colonies of social insects different individuals have different tasks. In a honeybee colony there is a single queen and thousands of workers that are actually sterile females, periodically there are also males, sometimes called 'drones'. Physically these three castes, queen, worker and drone, all look different. In ants, castes are generally much more distinct, and frequently greater in number. The 'workers' may even be physically distinguishable as foragers, nest attendants, major and minor soldiers. The presence in the fossil record of a worker ant of whatever kind would therefore indicate that it lived alongside other workers in the service of a queen. The problem is where to find a fossil ant. The answer may be hanging round your neck.

If you happen to like amber jewellery, you will know the attraction of a tiny piece of plant or perhaps a whole insect that was trapped in the golden resin flowing down a tree trunk one afternoon, 100 million years ago, and that still survives entombed, preserved and mineralized in perfect detail. There are various amber deposits from different past ages around the world, but a fossil worker ant in a piece of New Jersey amber, of that dinosaur-populated era the mid-Cretaceous, confirms that socially living ants did exist at least 100 million years ago and that consequently so did their nests. For termites you can double that figure and we have the fossil nests to prove it. Sandstone pillars in a location on the South Africa-Zimbabwe border have proven to be the remnants of termite mounds of the Early Jurassic, about 180 million years ago.[8] The preserved architectural detail shows these to have been complex structures reaching about 3m high. Termites had obviously been around for a while before these were built, indeed we have much simpler fossil termite nests from the Late Triassic, more than 200 million years ago. We can fairly conclude that there has been a long period of evolutionary time for social insects to create habitats through their building that other species could have become adapted to.

Over the millions of years, the influence of social insects on landscapes will have grown. Now, in tropical habitats, ants and termites can truly be said to be dominant among the animals. In parts of the Amazonian rainforest it has been estimated that ants and termites together represent about one third of the total animal biomass, that is, of the total weight of all animals, vertebrate and invertebrate. That is a lot of ants and termites when you consider that it would take about 30 million ants to balance the scales against a 140kg (309lb) jaguar.

I will give you results of just one study on a single species of termite (*Cubitermes*) in West Africa to suggest the extent to which social insect nests can influence their local ecology. *Cubitermes* build rather charming little nests of mud that look like fat-stemmed toadstools, about 35cm high that, if painted red and yellow would look just right alongside some plastic gnomes in an English suburban garden. This study showed that the number of ant species taking advantage of the space offered by *Cubitermes* nests was 151, eleven of them previously unknown to science.[9] Note, this is of ant species alone. However, these ant species appear to be non-specialists, opportunists that do not depend simply on termite nests for living space. However, millions of years of social insect evolution have given rise to many species that are indeed dependent upon the nests of termites and of ants to the exclusion of all other micro-habitats. Of the thirty or so different orders of insects, for example, at least ten have species that depend upon the nests of ants. Of beetles alone, there are probably hundreds of species, belonging to about thirty-five families, with ant nests as their natural habitat. Many butterflies, particularly those of the 'blue' and 'hairstreak' family, Lycaeneidae, have larval stages that are cared for by ants in their nests. In Europe the caterpillar of the large blue (*Maculinea arion*) initially feeds on wild thyme. However, when partly grown, an ant of the species *Myrmica sabuleti* will carry it inside its nest, apparently seduced by secretions from the caterpillar's body surface. Once inside, it repays the ant's hospitality by turning carnivore and eating the ant grubs. The full-grown caterpillar becomes a pupa within the protection of the nest, and from it emerges as a striking blue butterfly.

Now it is the turn of humans to attract species to the new habitats we have built. We are the dominant habitat-altering species; no other single species has altered the world so much by their building activity. Do the effects on habitats and biodiversity brought about by builders that have come before us tell us anything about what effects we may have? The initial impression is that, far from enhancing biodiversity, we are in the process of substantially diminishing it. The history of life on earth is recognized to have included five episodes of mass extinction. That is where between 10 and 40 per cent of species diversity has disappeared within a relatively short space of time, usually a few million years. The last of these was the extinction of the dinosaurs around 65 million years ago. I support the growing view among biologists that we are now in a sixth era of mass extinction and that it is caused by us. Notice, not 'about to enter—unless we are very careful' but 'in' a period of mass extinction. It began with the migration of *Homo sapiens* out of Africa (about 100,000 years ago, although this is a matter of some debate), completing our envelopment of the globe with the Maori colonization of New Zealand a mere seven or eight hundred years ago.

The accumulating evidence of our contribution to extinctions is that soon after our arrival in Europe and the Americas, large mammals became extinct. Large mammals never colonized New Zealand before humans introduced them; instead their grazing, browsing and predator roles were taken by large birds, including moas—species of flightless birds, two or more species towering above you or me, at 3m tall. Soon after human arrival, the moas became extinct, probably twelve species of them, along with several other species, including the largest eagle that ever flew, Haast's eagle (*Harpagornis moorei*). With an estimated wingspan of 2.6m, this was probably a predator of moas. If we compare ourselves with beavers, therefore, our capacity for habitat destruction is immensely greater. Nevertheless, we are undoubtedly creating new habitats and these provide potential living space for some species.

If a spider runs across the carpet as you are watching television or appears in the middle of the night stranded and sharply silhouetted against the white background of your bath, you may not just be startled but also upset at its invasion of your private space. However, this spider has almost certainly stumbled into your space in the house from one of its spaces, the roof, the wall cavity or the under floor. Do you ever consider what it is doing there? Spiders are after all the top predators of the invertebrate food chain in your house. There is a whole ecosystem 'below' them down to the tiny insects and mites that feed on your food crumbs, skin scales, and on the fungi growing on your house's damp patches.

Most of these house dwellers, like the ant species in *Cubitermes* mounds, are opportunists, but some of them are nevertheless heavily dependent upon us. The house martin (*Delichon urbica*), for example, is so called because its nests are found almost exclusively attached to the walls of houses, and rarely on the rocky cliffs where their ancestors nested. Some species, by occupying habitats made by us have extended their ranges dramatically. The brown rat (*Rattus norwegicus*) is so uniquely associated with human habitation around the world that, although scientifically termed the 'Norway' rat, it seems to have originated somewhere in Asia. The American cockroach *Periplaneta americana* seems to have been introduced into the New World from Africa, but what hotel in the world is now entirely safe from its intrusion? But are these species evolving under selection pressures exerted by us? This is an important question because we are a species that is exceptional in the rate at which it is altering its built environment. There is an Irish jig tune 'Cricket on the Hearth' that is a musical reference to the song of the male house cricket (*Acheta domesticus*), once a common resident in the thatched cottages of rural Ireland, but I don't suppose that there are many crickets now singing in the fitted kitchens of the smart bungalows now populating that land.

Electric and electronic gadgetry are habitats that proliferated in houses of the twentieth century. TV sets have been a potential household habitat for insects or spiders but, after a mere fifty years, the bulky cathode ray tube set is being replaced by the slim plasma screen set; far too little time for any creature to become specially adapted to its environment. The 'clothes moths', or at least their caterpillars that once fed on the woollen coats and carpets of our great-grandparents, have now retreated in the face of the human countermeasures of synthetic fibres and insecticides. For all our professed love of wildlife, we generally do not like to share our houses with it, particularly if it causes the least bit of damage. Nevertheless, however much human dwellings change, and let us be clear there are no guarantees of continued rapid technological change, the human house will remain a home, shaped to provide for our basic security and comfort. There will always be other species able to take advantage of that.

In Britain, the concrete ledges or exposed steel I-beams of high-rise buildings provide perching and nesting sites for feral pigeons, and peregrine falcons; our flat roof tops are the new cliff tops for nesting gulls. Spaces under houses, even in fully urban areas, are increasingly homes for that archetypal creature of the countryside, the red fox (*Vulpes vulpes*). So humans are creating new habitats by their building and, in spite of the rapid change of our way of life, other species continue to take advantage of these structures for their homes. Many of these also take advantage of other aspects of our biology, human domestic refuse in the case of urban foxes, but that is exactly the situation of beetles whose habitat is the rubbish dump chambers of leafcutter ants. How long will it take us to evolve the close mutualistic relationships that we see between, say, the goby and the shrimp, or between leafcutter ants and their fungi? Well, we have similar intimate relationships already: they are with our dogs and cats.

In the church of All Saints, in Little Shelford, Cambridgeshire, there is a fourteenth-century brass effigy of the knight Robert de Frevile in full armour, and of his lady, Clarice. They lie formally, side by side, staring expressionlessly at you, so it is with a shock of pleasure that you notice his right hand, ungloved, reaching across to gently hold her right hand, as her left hand touches her breast as if to muffle a 600-year-old catch of breath. But the brass tells us more of the domestic life of the de Freviles. His feet rest upon a fine hunting dog that gazes obediently up at its master, while at her feet, nestling in the folds of her long gown, are two lap dogs, the little bells on their collars once the everyday sound of the de Frevile home.

We like dogs. Some might say that we exploit dogs. Lap dogs or work dogs, we have bred them to our liking, but they are also exploiting us. We pay for their food and their vet bills. Initially, as a wild species, they had attributes that humans benefited from but we have now bred them to be more desirable to us. Their benefits are acknowledged as human companions, even as therapies for the mentally ill or the antisocial. In return we are securing their future as a species. As long as there are humans, dogs will not become extinct.

Somewhere, perhaps not very far from you, deep in the hillside is an elaborate and formidable bunker. In the event of any impending apocalypse, it will be the refuge of your *Emergency Regional Government*. Let me suggest to you who will get in, if and when the time comes: the ruling elite, some military, assorted partners, lovers, mistresses and children, and their dogs and cats. The spiders are probably already there.

So we live largely in a human built world, but so do termites live largely in a termite built world, and orb web spiders, largely in a web world. Animal builders are in a special relationship with the selection pressures that act upon them. They themselves create part of that environment. The final investigation of this chapter is to help us understand the evolutionary consequences of this, not simply for the evolution of the organisms that are builders, but upon all species that are in some way dependent upon them. This is the theme of 'niche construction', a term that is effectively synonymous with ecosystem engineering.

Both ecosystem engineering and niche construction explore the consequences of organisms modifying their environments through their activities. However, whereas the former emphasizes ecological effects, the latter stresses evolutionary ones. The term 'niche construction' has faced the same criticism as that of 'ecosystem engineering', that it is too broadly defined to be useful. Nevertheless, habitat modification by organisms is increasingly recognized as having far-reaching evolutionary consequences and the part of it that we are interested in, building behaviour, to be a very important aspect of it.

If the design of a spider's web is inherited from its parents, then those spiders must possess genes for features of that web building behaviour. I will talk more on the inheritance of building behaviour in Chapter 5, but suppose in the case of this web building, that there is inherited variation between individuals of a certain species in the number of radial threads ('spokes', if you like) in their orb webs. We can say that, at this gene location, there are alternative forms of the gene (referred to as *alleles*), for the number of radii in the web. Natural selection will act on this variation. So we might imagine that in a windy location, where stronger webs are more durable, webs with a greater number of radii will be more successful, so locally selecting against spiders carrying alleles for low radius number and therefore increasing the frequency of alleles for higher radius number. We are expressing evolutionary change as changes in allele frequency and explaining it in terms of straightforward natural selection.

The web of a spider, wind or no wind, is not going to last very long, much less even than the spider that spun it. But what about the casts made by an earthworm? There, as we discovered from Darwin's studies, the situation is very different. Young earthworms not only inherit from their ancestors the ability to burrow, but also a world altered by their burrowing. That is two pathways of inheritance, with the modified environment modifying the selection pressures that act upon the current earthworm generation.

Think back to the opening of this chapter: 'Wombats Detected from Space'. As we now see, not simply earthworms but termites, beavers and other builders, by their cumulative action over generations, can also substantially alter habitats, bequeathing those changes to their descendants. Attempts are now being made to predict how a niche-constructing species might influence its evolution through ecological inheritance. This is very difficult to do in a natural population of any organism, even one as relatively simple as an earthworm. This is where mathematical modelling by theoretical biologists can give us valuable indicators. Such a model has been devised and tested, and is explained in the interesting and challenging book, *Niche Construction*, written by John Odling-Smee, Kevin Laland and Marcus Feldman, published in 2003.[10]

Their model strips the evolutionary problem down to the simplest level possible. It asks the question: 'How can habitat-altering building behaviour over generations influence the evolution of some aspect of the organism in subsequent generations?' However, this is expressed in terms of allele frequencies, so the question is how do alleles at a single gene location responsible for building, influence allele frequencies at some other gene location over generations. Their model is a virtual creature with just two genes. You might well ask how such a highly simplified model, run on a computer using artificially chosen parameters of habitat change, can tell us anything. The best answer I can give is that you need to start somewhere and it is better to start simply. A theoretical model also has the merit that, although it is necessary to incorporate some assumptions because knowledge is incomplete, the assumptions are at least clearly stated. Later, additional evidence may invalidate certain assumptions, but we can then refine the model accordingly. Meanwhile the model shows outcomes of dynamic relationships that give us an idea of what we should be looking for in the real world.

So this two-gene model envisages that the organism's influence on changing the environment through building is affected by alleles at a one-gene location, which we shall call **E**. The environment, we are asked to imagine, contains some resource (**R**) which is altered by present and past levels of niche construction. **R** is, in other words, a function of the frequencies of alleles at gene location **E** over a number of recent generations. To make it less abstract, we can imagine a virtual earthworm where alleles at **E** determine the amount of its burrowing; this influences soil fertility, leading to plant growth and therefore the availability of more dead leaves, which provide food (resource **R**) on which earthworms can feed.

The amount of resource in the environment in turn determines the contribution made to the organism's fitness by alleles at a second gene location we will call **A**. Let's say, in the case of our virtual earthworm, that alleles at **A** influence a behaviour—the readiness to reach out of the burrow to grasp leaves. This completes feedback through the environment back to the builders. The influence on the habitat exerted by alleles at **E** alters the availability of **R**, which selects for certain types of allele at **A**, so changing allele frequencies at **A**, and, in our case, leading to worms that are more or less ready to reach out of their burrows than were their ancestors.

This model makes some significant predictions, but let's look at one, just by way of illustration: The time lag with which changes in allele frequency at the **E** location would impact on alleles at the **A** location. Suppose that the **E** alleles, which influence (through the worm's burrowing behaviour) the availability of the resource (**R**) (leaves), only begin to have a significant effect on **R** after a large number of generations of the burrowers. This results in a long time lag before the effects of alleles at **E** begin to impact on the alleles at **A**. This creates an evolutionary inertia which could, for example, see a particular allele at **A** continuing to decline in spite of the fact that an allele at **E** is altering the environment to its advantage. Less obviously, the feedback time lag could also generate evolutionary momentum. This would be seen if selection at the **E** location stopped or reversed. Because of the time lag on changing **R**, this does not prevent the resource continuing to accumulate for several generations, with the result that evolutionary change (in allele frequencies) at the **A** location persists for a while in the original direction in spite of the selection at **E** having stopped. In other words, the time lag produces a 'supertanker' effect on evolutionary direction: slow to get moving; slow to alter direction.

So, niche construction does appear to have important implications for evolution through two lines of inherited information: conventional genetic inheritance from parents, and through the inheritance of altered environments. But there is a third pathway for the transmission of information across generations, one that is immensely important for humans, the transmission of learned information. Humans store knowledge accumulated over generations, in libraries and databases, which is passed on to each new generation through an elaborate system of formal education. Is there anything equivalent in other species? The answer is that education in non-human animals is virtually absent. At a site in Guinea, West Africa, the chimpanzees crack oil-palm nuts by placing them on a stone anvil and striking them with a stone hammer. It takes at least three years for a young chimpanzee to learn how to do this, more to become skilled. To achieve this, he or she initially pays close attention to what the adults are doing and copies them. However, in spite of the fact that adults do leave hammers and anvils lying about, there is no convincing evidence that they are offering the youngsters a structured education.

There are of course a great number of species of animals that are able to learn. Any insect that builds a nest (ant, bee, wasp or termite) must learn how to get back to it. In a number of vertebrate animals, learned information passes from one generation to the next by youngsters copying the example of parents or other adults, but for the great majority of non-human animals there are really only two routes for the inheritance of influences that alter the environment: the genetic route, and the inheritance of an environment already altered (i.e., the ecological route).

For humans, the situation is dramatically different. It is now overwhelmingly through education (the cultural route) that we inherit our ability to change our environment. Students take degrees in engineering and science to be able to continue changing our world, and to discover new ways of doing it, which can be passed on to the next generation. So, what remaining importance do the other two routes of inheritance, genetic and environmental, continue to have for us?

The transmission of information encoded in DNA molecules is very effective when the message to be sent is relatively simple and when it is more or less equally applicable to the next generation as it was to this. It copes well with gradual change. Cultural transmission by education across the generations allows rates of change in behaviour that far exceed what is possible through genetic transmission; a radically new idea arising in one generation can become widely understood and applied in the next. However, even faster rates of social change than this are possible through individual learning. Individual learning becomes beneficial when environmental change is so rapid that the older generation have very little left to teach to the younger. Parents teach their children how to behave at table; the children teach the parents the special features on a mobile phone. Human inventiveness can bring about such rapid changes.

With rates of technological change appearing to become ever faster, parents may begin to wonder if they have a future role in raising children. However, we can exaggerate the rate at which we are altering the world and so altering the selection pressures acting upon us. Through the cultural transmission of acquired knowledge, we are the world's pre-eminent niche constructors. However, because we are already genetically and culturally adapted to live in a particular environment, we tend to create new environments that resemble past ones. We may have central heating with thermostatic control, but the room temperature we try to create probably resembles an environment that humans have preferred since the time they lived in simple shelters or natural caves. We saw how the mounds of *Armitermes* termites in the Northern Territories of Australia buffered the insects against the daily fluctuations in temperature and the seasonal fluctuations in food availability: it made termite lives more predictable, and their predictable way of life and relative population constancy brought greater stability to the whole habitat. The mounds of those termites are a conservative influence on change. Our niche construction, for all its innovation, has that aspect to it as well. However, as human history gets longer and our numbers increase, so we create more archaeology; that is, more altered habitat to pass to succeeding generations. These modifications may be to our current advantage, or merely haphazard outcomes of our varied activities, but for good or ill they will form part of the selection pressures acting upon our descendants.

3

You Don't Need Brains to be a Builder

It is a sphere composed of a few hundred stones cemented together, with a large circular hole at the bottom. The top of its dome bears seven or eight sturdy spikes, each a cairn of stones, larger ones at the base, the smallest at the tip creating a sharp point. The most distinctive architectural detail, the one that gives the name to the species that builds it, is the collar to the circular aperture. It is a pleated coronet constructed from particles too small to be distinguishable from the cement that binds them. The diameter of this whole dwelling, for that is what it is, is about 150 thousandths of a millimetre (i.e. micrometres, written μm). Smaller than the full stop at the end of this sentence, it is the portable home of *Difflugia coronata*, a species of amoeba (Figure 3.1).

As we established in Chapter 1 an amoeba is not an animal at all but a member of the kingdom Protista. Its single cell does everything an organism needs to do. It feeds, excretes, moves about and reproduces. It moves by flowing across the debris at the bottom of a pond or the like, sending out a 'pseudopod' (false foot) like a glacial flow, in one direction, and gathering up its irregular shape behind it. As

Figure 3.1. Amoeba case: a single-celled amoeba (*Difflugia coronata*), an organism with no nervous system, is able to build this intricate, portable sand grain house.

it moves, it engulfs tiny food particles and digests them, ejecting the remains in its wake. In this way the amoeba grows and periodically reproduces by dividing its body, and the nucleus which controls it, into two. What is probably less familiar is that an amoeba can also build a portable house that it carries round for protection, as a snail carries a shell, but only some species of amoeba do this; *Difflugia coronata* is one of them.

How does this single-celled creature build such an elegant house? Well, we don't really know. The only information we have at the moment is a description of what we can observe. An individual *Difflugia* flows around, carrying its case with it. While doing this, it not only engulfs food particles but also tiny sand grains that accumulate inside the amoeba as a large ball. When the time to reproduce arrives, the nucleus of the amoeba replicates its DNA to create two complete nuclei. The cytoplasm (the body material) then begins to divide, one nucleus going into each half, to form two independent organisms. One of these will inherit the existing house, but the other takes the ball of stones in its cytoplasm. As the two organisms are created, these stones move to the surface and arrange themselves as a new house.

That last sentence may sound pretty unsatisfactory. It is like a magic trick that leaves you wanting to know how it was done rather than simply enjoying the moment, but we simply don't have the information. You may also have another feeling of dissatisfaction. An amoeba is a single cell. Isn't this therefore a story about cell biology not about behaviour and so has no business in this book? Well, I think case building by *Difflugia* makes a fundamentally important point about building behaviour. You don't need brains to be a builder.

I hope that we can agree that an amoeba, with or without a portable case, has behaviour. When it sends out a pseudopod in one direction rather than another it is showing behaviour based on some decision. Even if you wish to say that the direction chosen is random, still there must have been some instruction generated within the organism to do something rather than do nothing, and indeed to engulf this particle but not that. I've no idea how that can be done, but let's imagine some essentially mechanical process. Some sand grains are too small to be grasped and carried into the cytoplasm, some too large. The result is selection. But clearly the house-building amoeba is making more decisions than that. It needs to ensure that it has enough sand grains to build a new house. Maybe it also avoids having too many. It needs very small particles as well as large ones. Can it tell as it goes along that, for example, it is still short of very small ones? Finally, what about the construction process? Clearly, there is a process that moves particles to the appropriate places and assembles them in a very special way. That process requires some equipment to manipulate the building material. We use hands to manipulate bricks and mortar. *Difflugia coronata* uses some intra-cellular equipment. This amoeba does show building behaviour, for which it needs equipment to decide what to do and equipment to carry out the building instructions. If a single celled organism can do those things and produce such a seemingly sophisticated result, then there must be ways of building that are very much more simple than the methods that we use: simpler in both decision making and building equipment. This chapter is an exploration of simple ways to build elaborate structures.

Animals, unlike the Protista, are all multi-cellular, with cells specialized to form different tissues, brain, nervous system and muscle. Decision-making and communication are the jobs of the nervous system. Movement is achieved by muscle contraction which often, although not always, operates a system of levers that we call the skeleton. Dissect out a human brain and it will weigh about 1,400g, and have a volume of around 1,450cu cm. The volume of the total central nervous system of one of the largest of the orb web spinning spiders is 303×10^6 cubic micrometres (cu μm). That gives the average human around 5 million times the brain size for decision and instruction that a spider has. From this comparison we should have certain expectations about how building behaviour in non-human animals, particularly among the invertebrates, should be organized.

These expectations can be expressed as three predictions. The first is that animal building behaviour will be kept simple. The process of natural selection will favour building routines that have a limited repertoire of behaviours, each of which is rather stereotyped and invariant. This does bring with it certain difficulties, in particular it cuts down the capacity for individuals to improve building skills by learning, but learning would require additional brain cells and circuitry. This leads us to a second prediction: that animals will tend to use standardized materials because, if the building materials are predictable in character, then the handling process can be invariant, offering the prospect of stereotyped repetitive building routines.

Simplicity in the construction process is not without benefits even to us; it can save on time and effort, and that in our currency means cash. Building a brick wall is an illustration of exactly the behaviour we are predicting in other animals. If a truck dumped a load of bricks in front of the cave of a Neolithic family, I could describe to them over the phone how to make a wall. We could leave the mortar out for the sake of simplicity, but they would quickly grasp the principle. However, there is one obvious stumbling block, if you will excuse the building pun: that is the problem of the cave folk getting their wall started. It would be easy to describe to them the rules for adding the next brick to a wall that has already been started, and each successive brick is subject to the same, simple set of rules and can be handled in the same way, but what about getting the wall started? Is, for example, the cave mouth flat or uneven? Maybe there is a big tree growing there. These are just the kinds of contingencies that make some kind of flexibility in the behaviour desirable. Once a structure has been started, then new structure is added to existing structure. This, the builder has control over, so stereotypy works. This brings me to my third prediction: getting started should be the part of the building sequence where the builder will exhibit its most variable and complex behaviour.

Now, what about building equipment? What predictions can I make about the personal apparatus (the design of skeleton and muscle) which simple animals should use in their building? Building behaviour evolved from behaviour that was originally nothing to do with building. There is no controversy in saying this. It is the way that anything, behavioural, physiological or anatomical evolves. No committee sat round a table, deliberating and designing from scratch: it arose from the modification of something that in previous generations worked in a rather different way and had a rather different function. So, what did building anatomy evolve from?

One of the pleasantly predictable attributes of TV sci-fi series, established by *Startrek* and accepted ever since, is the appearance of aliens. From whatever strange new civilization, from whatever part of the galaxy they may hail, vary as they may in the number or colour of scales and lumps on their heads, they will still have two eyes, two arms and walk about on two legs. I very well see the economic argument of the studios that dressing up some workaday actors in funny head costumes is a cheap way to create exotic life forms, but I like to think that this may well be the reality, that in some parts of the galaxy there really are exotic metropolises, streets thronged with workaday commuters who, but for a few lumps and scales, look not unlike you or me. There is a biological justification for this. It is the argument that there are only a few good solutions to any problem and that in our body design we have incorporated a number of good solutions.

I want to apply this argument in a somewhat speculative way to the evolution of the anatomy of builders. In other words there are only a limited number of bits of the body of any organism—human, fish or spider—which are suitable for modification as building equipment, and only a limited number of ways that they can be effectively modified. This may be a rather weak argument, but it is simply a hypothesis with some clear predictions that we can test by looking at the evidence. So here are three predictions it yields on the nature of the anatomy of builders.

The first is that the degree of specialization in the design of building anatomy will depend on the extent to which it still retains its original function. The second, based upon the argument that there are only a limited number of good solutions, is that we should expect that parts of the body used for building will have similar origins in vertebrates or invertebrates—in fish, spider, wombat or wasp.

Building requires some manipulative skill, so anatomical adaptations should reflect this. However, there is another aspect of building, and that is power. My third prediction is that adaptations in anatomy will be evident where building requires power and the most obvious place to look for it will be among the burrowers, since they need considerable power for digging. For the moment I am going to leave this as a bald, unsupported assertion, but I will come back to it later.

These predictions probably sound a bit abstract, so here are some illustrative examples. The beaks of swallows and martins are an example of the first prediction: that degree of specialization reflects degree of use. The sand martin (*Riparia riparia*) is a burrower, using both beak and feet to create its nest tunnel and cavity. The purple martin (*Progne subis*) utilizes natural cavities or, nowadays, nest boxes, only using its beak to gather a few scraps of building material. Barn swallows (*Hirundo rustica*), cliff swallows (*Petrochelidon pyrrhonota*) and house martins (*Delichon urbica*) use their beaks to gather and fit together mud pellets to build up nests of varying complexity (page 126, Figure 5.2).[1] With that in mind, look at Figure 3.2 which shows the heads of these four species and, without looking at the caption, decide which is the burrower. Well, they all look pretty much the same. There is no obvious specialization in their beaks for their particular form of building. So what are the beaks for? Well, obviously, in addition to building, they are used to capture and handle prey, and in the case of all these species, to catch insects while on the wing.

Let's do a second test, this time on the feeding habits of birds. This time we'll use a group of birds found only on the Hawaiian islands: the Honeycreepers. These birds are an even more dramatic, albeit less well-publicized example of what Charles Darwin found in the finches of the Galapagos Islands: the diversification from a single pioneer species of a clutch of species adapted to different local habitats. The total area of the eight Hawaiian islands is somewhat larger than that of the Galapagos, but with more luxuriant vegetation and mountain peaks of over 4,000m, compared with the 1,700 maximum in the Galapagos. This has contributed to the adaptations shown by this

Figure 3.2. Similar beaks, dissimilar nests: (d) the sand martin, digs a burrow nest; the other three species, (a) barn swallow, (b) chift swallow and (c) house martin, all construct nests of mud pellets.

bird group. Now look at the head profiles of some of these Honeycreepers shown in Figure 3.3 and, again before reading the caption, decide what sort of food is eaten by each species.

After checking the caption, you are probably rather pleased with your success. But now try to decide what sort of nest each of these species builds. You are forced to guess, aren't you? The beaks provide no information. In fact, across the group, the nests are generally cup-shaped, placed in the branches of trees. They are made of a variety of mostly plant materials, but essentially they are not that different between species. So, in spite of the fact that all these species depend utterly upon their beaks for nest construction, there is no way of deducing anything about their nest building from their beak design.

One of the most attractive birds that comes to the thistle seed feeder in my garden is the goldfinch (*Carduelis carduelis*). These are

(**a**) Akialoa

(**b**) Ula-ai-hawane

(**c**) Kona grosbeak

(**d**) Common amakihi

small, neat finches with scarlet faces framed by a black cap and white cheeks. In winter they come with the morning light and only leave at dusk. For a large part of every day a goldfinch is using its sharp triangular beak for splitting open small seeds. So, how much time does a goldfinch spend building a nest? There is no precise information on this, but let's say three hours a day for four days. So, about twelve hours a year a goldfinch uses its beak for nest building, and that's only the female; the male generally provides no assistance. The nest of the goldfinch is a beautifully neat cup made of moss, rootlets and plant down, but for all that, the bird's beak is adapted and used almost wholly for feeding. Birds as a group uphold our first prediction about the anatomy of building. Its specialization reflects the proportion of time it is used for building.

Let's consider the first of the predictions on building behaviour; that it will be selected to favour a limited repertoire and stereotyped, repetitive actions. Evidence for this is offered by both brain anatomy and behaviour itself. There is almost no evidence that building behaviour is associated with specialized brain areas. I have to admit this could be due to lack of curiosity from researchers, but if we take the birds again as an example, what does the evidence show? Birds have a large brain relative, let's say to spiders, and build elegant nests, yet we know of no specialized nest-building areas in their brains. This is in contrast to the known situation for birdsong. For song production there are a small number of very obvious 'nerve centres' (termed *nuclei*) that are concerned with the learning and production of song, that is discrete localized concentrations of nerve cells linked by highways of parallel nerve fibres—a dedicated song production system. To the initial surprise of biologists, some of these birdsong nuclei are also dynamic structures, shrinking in size as one breeding

←

Figure 3.3. Beaks indicate feeding habits: Hawaiian honey creeper species have markedly different beaks adapted for their specialized diets. The main foods of the four species are: (a) nectar, (b) small soft fruits, (c) tough seeds, (d) small insects. (Note: (a),(b) and (c) may now all be extinct.)

season ends, and enlarging with the production of new brain cells as a new breeding season approaches.

Specialized brain structures have also been found in some birds that are concerned with food finding. This is illustrated by the coal tits (*Parus ater*) that visit the sunflower seed bird feeder in my garden. At first light on a winter morning the first coal tit will arrive, grab a sunflower seed and fly off, probably over the stone wall and into the lane behind. Within the minute it is back to grab a second seed, repeating the process without interruption five or six times. What it is doing is hiding the sunflower seeds in cracks in the wall. In the evening it will return to recover them for supper before going to roost. When summer comes, the occasional stunted sunflower sprouting from the stonework of my house is a reminder of this characteristic coal tit behaviour.

Blue tits (*Parus caeruleus*) and great tits (*Parus major*) also take sunflower seeds from my feeder but they just take them into the bush close to the feeder, immediately stab their way through the seed coat and eat the contents. There is an area of the brain in birds (and indeed in mammals) known as the hippocampus (the same as the scientific name for a sea horse because of its shape, and derived from the Greek *hippo*—horse, *kampos*—sea monster). This paired structure is known to be associated with spatial learning. In the coal tit this is significantly larger relative to overall brain size than it is in blue and great tits, an adaptation to recovering hidden food items. Nest building in birds apparently requires no such specialized brain regions.

We have too little information on the workings of invertebrate brains to make the same comparisons as in the case of birds. Do caterpillars that make elaborate cocoons have special brain structures that are absent in species that build no cocoons, or even have larger brains overall? We don't yet know, but I think it is fair to conclude on present evidence that building behaviour is undemanding on nerve circuitry.

What about evidence from the building behaviour itself: is it simple and stereotyped, as predicted? There is a type of small tadpole-like creature (*Oikopleura dioica*), no more than a few millemetres long, that drifts in the plankton inside a mucus capsule of its own making. In some species these capsules can be about 15mm long and in numbers as dense as a snowstorm. These mucus structures are very interesting because they are designed both as a house and for food gathering. The organisms that build them are chordates, the phylum that includes all the vertebrates. However, they belong to the Appendicularia or Larvacea, which do not have a true backbone and so, on the scale of vertebrate evolution as well as of body complexity, are simple creatures. *Oikopleura dioica* is shown in its house in Figure 3.4.

Oikopleura dioica drives water through its house by lashing its tail. This brings in the food particles and dissolved oxygen that it needs to survive. The water enters through a pair of inlet funnels within each of which is set a mucus net with a regular mesh of about $30 \times 100\mu m$ (thousandths of a millimetre). These nets serve as a barrier to prevent larger lumps of material and also unwanted creatures from entering. The water then travels through a pair of filter nets, each of which is in the form of a sandwich of upper and lower filter nets with a mesh of only $0.3 \times 0.3\mu m$ (micrometres), between which is a large meshed scaffolding net for support. The tadpole creature itself harvests the particles of food from the filter nets and the filtered water passes out of the house and back into the ocean.

So how does *Oikopleura dioica* make this elaborate and delicate structure? It does very much what it does when feeding: it just lashes its tail. Well, it's slightly more complicated than that, but not much.[2] First it secretes from glands on its head a mucus helmet, which it enlarges initially with blows of its head generated by vigorous lashing of the tail. In this way the capsule becomes big enough for the tadpole to slip its tail inside. It can now complete the inflation of the capsule through the direct effect of tail lashing (Figure 3.4).

But how are the barrier nets and the filter nets built? Well, they just appear. If you feel a bit cheated by this answer, you shouldn't because it contains an important revelation. The answer to the question 'How can a complex structure be built by a creature with a small brain,

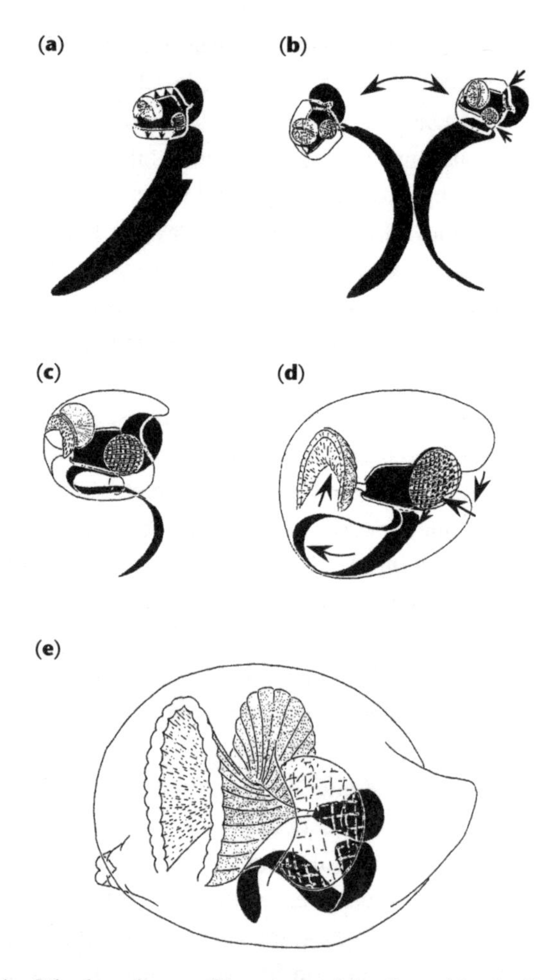

Figure 3.4. ***Oikopleura*** **house: Clever materials allow this planktonic creature to build a complex home with simple behaviour—after initially secreting a mucus capsule around its head (a), the animal then expands it with vigorous head movements (b); this permits expansion of the house to full size, using simple tail lashing (c–e).**

Adapted from Flood, P. R. 1994. Appendicularian—Architectural wonders of the sea. In *Evolution of Natural Structures* (Proceedings of the 3rd International Symposium Sonderforschungsbereich 230), pp. 151–56. Universitat Stuttgart and Universitat Tubingen very simple behaviour and unspecialized anatomy? is 'By using clever materials.'

I'll say more about the significance of clever materials in prey capture in Chapter 6, but the important thing to note here is that *Oikopleura* builds its house out of material that it manufactures itself. Even a simple animal with a small brain and limited repertoire of behaviour has elaborate molecular biology and biochemistry. This may have allowed *Oikopleura* to evolve highly specialized secretions, including ones used for building. Self-secretion of building materials offers important benefits for building behaviour. The first is that it standardizes the materials used in construction; the gland itself provides the quality control for the composition of the building material. Second, there is no need to collect material for construction. When the construction phase is ready to begin, the material is already there.

If standardization of materials is so important to animal builders in allowing them to keep behaviour simple, we would expect to see those animals that depend upon collecting rather than secreting their building materials, also to find ways of standardizing them. There is abundant evidence that they do, and they achieve it in one of two ways: by selecting from the world about them only bits of a certain kind, or by selecting a raw material that is standardized through a manufacturing process. Both these are illustrated where the standard building unit is a brick.

The caddis larva *Silo pallipes* exemplifies the first method. It initially scratches the ground with its front legs until it detects sand grains. This is followed by the picking up of a particle within a wide range of sizes and shapes—apparently anything that is neither too large nor too small to handle. The particle is then held in all the legs and repeatedly turned round and over and may well be rejected. A sand grain passing this test is held to the anterior rim of the house and tried in different positions in different orientations with all the legs, while the mouthparts seem to assess the closeness of the fit. Some particles fail this test and are rejected. The larva is standardizing the materials used by the application of three apparently fixed tests. However, the behaviour of handling and manipulation does show a degree of flexibility that characterizes a dry stone wall builder. Just how flexible, would be a useful target for future research.

I don't think it would be unfair to say that, whereas the builder of a dry stone wall is regarded as a craftsman, a bricklayer is regarded as less skilled, the reason being that the bricklayer has the benefit of building blocks manufactured to a uniform standard. (There is actually a standard metric brick. It has the dimensions 215 × 102.5 × 65mm; allowing for 10mm of cement bonding, this makes a filled size of 225 × 112.5 × 75mm, a ratio of 6:3:2). Many species of caddis larvae build with bricks, or to be more accurate leaf panels, cut by the individual builder to its own specifications. The caddis larva *Lepidostoma hirtum* cuts more or less square panels that are fitted together to create a four-sided box girder, one row of panels for each side. The panels in neighbouring sides are half a panel length out of phase, the roof projecting at the front of the house half a panel length in front of both sides, which project another half-panel length in front of the floor. This gives the head of the larva some protection from above and the sides, and its legs some freedom to walk as it leans out of the front of its case.

The caddis larva, after cutting a new panel, applies a very simple rule on where to place it. It will be to the least projecting of the four sides. This is of course the 'floor', which brings its front edge to the level of the roof. Applying the fitting rule, the position of the next panel is a 'toss up' between the original left and right sides, but whichever it is, let's say the right side, becomes the new roof, the original roof and floor become the sides and the original left side becomes the floor. The larva rotates inside its case, responding to the new house configuration. By manufacturing its own bricks, this species reduces the complexity of the fitting process compared with the sand grain manipulation of *Silo pallipes*, although some complexity is added by the behaviour needed to manufacture the leaf panels in the first place.

The importance of the brick in making building behaviour easy is in fact so great that there are species that secrete their own building materials in the form of bricks. This has evolved independently in a number of species, and the building blocks are dung pellets.

As a schoolboy I used to hunt for caterpillars; privet hawk moth caterpillars (*Sphinx ligustri*) were the special prize. The bush where they could be found each summer overhung a path so it was there that I looked for really large faecal pellets, uniform as little beer barrels. The faecal pellets of caterpillars are often quite dry and hard, truly brick-like. There is a family of moths, the 'bagworms' (Psychidae), so called because the caterpillars build themselves portable cases like caddis larvae, although bagworms live in trees. I have collected these cases from various locations around the world and, many years after collecting one from Malaysia, I decided to examine the composition of its smooth, narrow tapered shape under the microscope. It was a revelation. The surface, which appeared to the naked eye to be unremarkable, turned out to be a regular spiral of tiny faecal pellets. One species of caterpillar from Australia has, however, evolved a modification of its faeces to produce not faecal bricks but faecal beams. Faeces in this species are voided not as single pellets but as rods composed of three or four pellets welded together end to end. The caterpillar then builds a shelter for itself on a leaf by silking the rods together in an arrangement of uprights and roof beams, over which it spreads a silk screen to create an enclosed tube. These two self-secreted materials, silk and faeces, confine the building behaviour to construction alone, using highly standardized materials.

Silk, like mucus, is another plastic (i.e. malleable), self-secreted material. Both are produced in a viscous form and, in the case of silk, drawn out into a thread. There are also some plastic collected materials, the most widespread of which is mud. Mud is used as a major nest component by about 5 per cent of bird species and by several insects, notably some solitary wasps that make cells out of mud in which to raise their larvae. To most of us, mud is just earth after too much rain, but anyone who makes pottery specifically chooses clay because of its fine particles, and then carefully adjusts its moisture content so that it can be readily moulded into a particular shape before it hardens. For a potter, 'mud' does not adequately describe the raw materials of pot-making. We should equally expect an insect pot builder to be particular about its mud.

In some mud building wasps the material is collected directly, but little is known about either the particle sizes or water content of it. In species such as the potter wasp *Zeta abdominale*, however, the mud is manufactured by the wasp carrying a crop full of water to a chosen patch of earth which is then regurgitated to create the mud. This obviously provides an opportunity for the wasp to control both type of soil and the quantity of water used in the preparation of the mud—an opportunity for producing a standard mud, adapted for building.

I went through a phase of trying to make pottery myself but, bowing to realism, sublimated it into collecting pots instead. But it introduced me to the pot-making techniques of 'throwing', 'slab' potting and 'pellet' potting. Throwing a pot is of course what potters do on a wheel. Slab pots are made by rolling out sheets of clay like pastry, cutting your shapes—to make a box, say—and sticking them together. Pellet pots are made by repeatedly pressing small balls of soft clay together to gradually build up the shape of the pot; this is the type of pot-making, with its manageable loads and standard units, that is used by animal builders.

The problem for potters making pellet pots, slab pots, or indeed putting a handle on a thrown pot, is cracks. Clay shrinks as it dries; that is why the mud at the bottom of a dried-out reservoir is cracked. In my potting days, sometimes, in attaching the handle to a mug, I was inadvertently pressing together two pieces of clay of slightly different moisture content. Pot and handle shrank to a different extent, and a crack developed before the mug was even placed in the kiln, where the handle then fell off. This problem faces all potters (human or animal) but there is a solution to it: the property of thixotropy, where a material can be stable when at rest, but fluid when mechanically agitated.

When you stand on the beach a few metres from the water's edge and trample your feet rapidly, the sand seems to turn to liquid and your feet begin to sink in; it is exhibiting thixotropy. When the solitary mud dauber wasp *Trigonopsis* returns to a partially built cell carrying a lump of mud in its mandibles, it holds the pellet to the cell, at the same time injecting a little more water into the pellet from its crop. The wasp then emits a soft buzzing sound, the vibrations of which liquefy the mud, at the same time spreading the mud backwards to build up the cell. Swallows and martins also appear to exploit mud's thixotropic properties. They collect mud pellets from puddles around houses and farms and, with rapid dabbling movements of the beak, weld each new pellet to the growing nest. The word 'weld' seems appropriate here because the effect of the dabbling is to make the water in the mud more mobile, allowing it to flow out of the pellet and into the dryer nest structure. The junctions between the pellet and the nest then share the water, vibrate together and, as the vibration stops, have a common consistency.

But do neighbouring pellets in a nest have a similar composition in terms of particle size? Is there a standard mud used by all barn swallows? There is tantalizing evidence that this is the case. One study compared the composition of the mud used in nests of barn swallows and cliff swallows nesting in the same locality, each of course nesting in the place their name suggests; the former making shallow bracket nests, and the latter making deep bowls with entrance spouts (Figure 5.1). The data from the study seem to indicate that the mud of barn swallow nests has less sand and more fine-grained silt than does the mud of cliff swallow nests. I would like to see this work repeated. If confirmed, we would then need to find out what the two mud types were adapted for—ease of manipulation, strength of the completed structure? A nice little research project there. But, as far as the here-and-now goes, this is just one small example in the general body of evidence for the standardization of building materials, an important contributing factor to keeping animal building behaviour simple.

Mud builders show that building is more than putting the bits together; the bits must be made to stay together. This may need special 'fastening together' behaviour—vibrating the material in the case of mud builders. A rock climber will use a special knot to fasten two ropes together, a double fisherman's knot for example. A rope has no natural tendency to attach itself to another rope, so complex behaviour is required for fastenings when using this material. The two materials that make up a Velcro fastening, on the other hand, only need to be pressed together, to stay together: complex materials, simple behaviour. The question is, do animals exploit the tendency of some materials to fasten themselves together in order to simplify fastening behaviour?

Using an adhesive or glue places more responsibility on the material than on the behaviour in fastening two pieces of building material together. The chimney swift (*Chaetura pelagica*), which attaches its diminutive bracket nest to the inside walls of buildings, uses salivary mucus to glue together the short pieces of stick that are the sole other component of its nest. A caddis larva uses silk as the glue that fastens the sand grains or leaf pieces together, spinning a double strand of silk from its mouthparts back and forth to attach each new piece to the front of its portable house. This quite simple, repetitive behaviour is not quite as simple as pushing together the two materials of a Velcro fastening. However, there are animals that do in fact use the Velcro principle. Its use in bird nests is actually quite widespread, found in at least twenty-five families. A good example of the use of Velcro fastenings is the nest of the long-tailed tit (*Aegithalos caudatus*). Its nest is a flexible bag with a small, round entrance near the top, usually located low down in a gorse or bramble bush. It is stuffed like a duvet with feathers for insulation, commonly 2,000 or more of them, and covered over the outside with hundreds of flakes of pale lichen. The bag itself, what holds the whole nest together, is a Velcro fabric, the two components of which are certain small-leaved mosses and fluffy, silk spider egg cocoons; the fine leaves of the mosses are selected to provide the 'hooks', and the spider cocoons provide the 'loops'.

The nest of the long-tailed tit is neat and elegant. It is an assembly of about 6,000 separate pieces, yet they are of only four distinct materials: lichen, feathers, moss and spider egg cocoons. The Nobel laureate, Niko Tinbergen, writing of these nests in 1953 said, 'the most amazing thing about it [the construction behaviour] is, in my opinion that so few, so simple and so rigid movements together lead to the construction of so superb a result'. The bird species that use this combination of plant materials and spider silk as a Velcro are taking advantage of a simple method of nest construction. If you are saying to yourself, 'surely spider's web silk could be used as a sort of sticky tape', then it seems that you are wrong. Some spider's webs are indeed coated with sticky droplets to capture prey, but these soon dry out. spider's web silk is used in nest building by many kinds of small birds as well as or instead of cocoon silk, but always, it seems, as Velcro loops, not as sticky tape. Silk is an immensely important building material. Spiders, caterpillars and caddis larvae, as well as some other insects, secrete it for their own use, but we now see that, second hand, it is an essential component of the nests of a large number of small birds: as a component of Velcro or, as we saw in Chapter 1, to make pop rivets to fasten the nest of the little spiderhunter, and by some bird species, to make silk stitches.

Stitching is a fastening technique that requires a certain level of skill. Maybe you have actually used one of those neat, 'first aid for clothes' kits of needle and thread found in hotel bathrooms along with a miscellany of sachets and bottles. In re-attaching an errant button you face the behavioural problem that, having pushed the needle through the cloth and the buttonhole, you must then let go, pick up the needle on the other side and drive it back through again. That may not seem very demanding but very few other animals can do this; and virtually all of them are birds. The appropriately named tailor bird (*Orthotomus sutorius*) usually combines silk and plant down to make short lengths of yarn to stitch together neighbouring living leaves to make a hanging purse which is then filled with fine grasses to form a nest. The 'needle' is the beak. To link the two leaves together, the bird drives the thread of yarn through one leaf with the beak, lets go, picks up the thread on the other side and drives it through the other leaf. Two or more stitches with the same thread may be done in sequence and more than one thread is needed to secure two leaves together. The bird is showing a significant level of complexity in its fastening behaviour, but this is certainly exceeded by another kind of bird, the so-called weavers.

The 'weaving' that characterizes the weaver birds is not generally the regular over-and-under, warp and weft by which we create cloth on a loom. Nevertheless, these birds address the same problem that we do with ropes: how to get two or more long strands of material, with no affinity to adhere to each other, to stay together. There is of course for both the birds and us the problem of how to get the strands in the first place. We spin yarn from plant fibres, but for the birds it is much easier than that. They exploit a design feature exhibited by the leaves of monocotyledonous plants or monocots. Grasses, palms and lilies are all monocots and the veins in their leaves lie parallel, running from end to end. Broad-leaved trees, roses, indeed the majority of plants, are dicots. The veins in their leaves are branching, even lace-like. If I plant the seed of a dicot, say of a sycamore or a carrot seed, the first leaves to push their way out of the ground are a pair of simple green paddles, between which the characteristic leaves soon emerge. These embryonic leaves are the cotyledons, two of them, hence *dicot*. A grass seed embryo has only one, hence *monocot*.

Botanical digression over, let's get back to the weaver birds. How does a village weaver *Ploceus cucullatus* manufacture a strand of building material? It lands near the base of a leaf of elephant grass and cuts into it with its beak, severing a few veins. Holding the cut end, it now flies off, pulling a long narrow strip of leaf that appears behind it as a tear travels up the leaf between the veins—a very simple procedure.

The problem now facing the bird is much more severe. How can these essentially linear building units be made into a three-dimensional hanging basket? Even assuming that the nest is already started, this has to involve the insertion of the thread through the nest fabric, and its recovery, followed by repeated insertion and recovery until the whole strand is now part of the fabric. This necessitates careful and constant beak–eye coordination and the use of recognizable stitches. A classic study of the 1960s discovered that weaver species use a variety of fastenings: ones that would be familiar to any sailor, including spiral binding, half hitches, simple overhand knots and even slipknots, as well as more or less regular over-and-under weaving.[3]

In fact, not one but two groups of birds have independently evolved nest 'weaving' behaviour. The village weaver is a member of a sub-family of birds, the Ploceinae, found in Africa and Asia. They belong to the same family as the ubiquitous house sparrow. These 'Old World' (Europe, Africa and Asia) weavers are birds of 40–60g, and the males do most of the building. Quite independently, in the New World oropendolas and caciques—birds related to finches and the numerous New World tanagers—also make hanging-basket nests, with the females doing most of the building. The Montezuma oropendola (*Gymnostinops montezuma*) manufactures plant strips from parallel-veined leaves in exactly the same way as a village weaver—cut into the leaf, hold the cut end and fly off. Our knowledge of the fastening used by this group is limited but we do know that it includes spiral binding and half hitches. Although oropendolas and caciques are generally much bigger than Old World weavers (a Montezuma oropendola female is 225g), their beaks are long and sharply pointed. Much more suited to weaving, you would imagine, than the short triangular beaks of their Old World counterparts.

The fastening techniques of these two groups of birds provide the most powerful example used so far in this chapter of complexity in construction behaviour. As the materials will not fasten themselves, complex behaviour is needed to unite them. Following and retrieving the same strand of material during spiral binding or tying a knot would appear to be a skilled task, bearing in mind that the beak is the only instrument used in the manipulation. If this is really so, then we might expect evidence that practice makes perfect. Do weaver birds, with practice, make better nests? Once again, the evidence we have is extraordinarily meagre, but it is at least from actual experiment rather than simply observation.

In an experiment conducted over forty years ago, young male village weavers were, in their first year, either given (controls) or not given (experimentals) experience of handling fresh green building materials. In a comparison of the skills of the yearlings of both groups, the experimental group males were unable to weave a single strip in the first week that they were supplied with reed grass. Even after three weeks of practice their success rate was only 26 per cent compared with 62 per cent for the experienced controls. Even in tearing strips of building material from the leaves supplied, the experimental birds were more inept. I continue to be amazed at the lack of studies on the learning of nest building skills by birds, and by the widespread assumption that nest building behaviour is all genetically determined and so requires no learning. This may prove to be largely true, but we need to know. In the meantime it is interesting and valuable to see that what appears to be one of the most difficult fastening problems facing builders does also provide evidence of skill learning.

By 'skill', I mean something more than an efficient mechanical performance, rather a fluency that comes from repeated practice with the building material. You can tie your shoe laces in a double bow; so can I. However, although we may achieve the same end result, I bet that your finger movements and grips differ from mine. We have our own personal idiosyncrasies. Do individual weavers have personal mannerisms and knacks in nest building, suggesting individualized skill development? That's something I'm looking into.

Weaving a strip of material into the nest may be difficult, but I have not yet dealt with what, for these birds, is probably the most difficult part: getting the nest started. The nests of most weaver species hang from a fine twig at the end of a branch, a protection against tree-climbing predators such as snakes. The first challenge for the bird is to fasten the initial long strip of material to the twig using only the beak for manipulation, albeit with the aid of the feet to hold it in position.

My third behavioural prediction at the start of the chapter was that we should expect the least stereotyped and most varied building behaviour to be needed for getting structures started. The problem for the weaver illustrates why. Not only does a strip of vegetation show no intrinsic aptitude for adhering to the twig, but the twigs at the ends of a branch will vary in their number and configuration. These make stereotypy in the building routine difficult.

Once the nest is established at the end of a branch, each new piece can be added to pieces already fitted. The building process is now entirely within the control of the builder; simple and repetitive behaviour is therefore more likely to work. This problem and evidence for my prediction have hardly been looked at in any systematic way. However, the problem was identified back in the 1950s in a scientific paper that pointed out the variation in the attachment of the nests of a number of species of tree-nesting birds. For example, the northern or 'Baltimore' oriole (*Icterus galbula*), a close relative of the oropendolas, may suspend its nest from above or attach it both above and below, securing it to varying numbers of branches—two, three or four. In other respects the nest of the Baltimore oriole is a predictable and species-typical deep pouch of woven plant fibres, suggesting less variable construction behaviour once the attachment has been secured.

The nest building of the village weaver further illustrates the transition from improvisation to predictability in the building behaviour. The domed nest is always suspended from the end of a branch. Inexperienced birds have great difficulty in getting the first few grass strips. David Attenborough's TV wildlife series *Trials of Life*, for which I was a science adviser, had a long sequence of an inexperienced male weaver struggling to bind the first three or four strips of grass on to the dangling twigs. Just when it looked as if a breakthrough was about to be made, the whole nest foundation detached itself. Upside down, with ineffectual wings half extended, the hapless weaver disappeared through the bottom of the TV screen, followed by the failed nest attachment to which it still clung.

What this weaver needed to achieve was a vertical ring woven of grass strips hanging beneath the suspension. Standing within this ring, facing always in the one direction, the reach of the bird then defines the dimensions of the nest. Reaching forwards, the tip of its beak describes an arc that defines the size and shape of the nest cavity; reaching up and leaning further and further back, the bird defines with the tip of its beak the curved profile of the porch that protects the entrance to the nest cavity. The bird is using the reach of its body as a template.

Using the body as a template is a widespread device to help simplify the building procedure. The beautiful circular section of a caddis case is easy for the larva to achieve. Applying the rule of adding the next sand grain to the least projecting point on the anterior rim, all the larva has to do is to reach out, holding the sand grain in its legs, and attach it with silk. The dimensions of its legs define the attachment point, which is the same distance whichever direction it faces, hence a perfectly circular tube. We will come across templates again, but they are another way of keeping building behaviour simple.

Incidentally, honeybees use their bodies to create the wax cylinders around them that will form the cells of the honeycomb. But, you may be protesting, surely honeybees make those wonderfully perfect hexagons, an example of their masterful construction skills. Well, it seems they don't, and it isn't. What they do is form a cluster on the comb, inside which some bees start to build cylinders. At the same time the cluster heats itself up by the 'shivering' of their collective flight muscles. The semi-molten wax cylinders then just flow together and, like the clusters of soap bubbles in your bath, create a beautiful geometry.[4] The building of hexagonal comb cells by wasps out of paper pulp? No molten magic there; it does require more control in construction process, although we have little information on how.

It is time to review my building behaviour predictions. Animals do show evidence that natural selection has favoured simple building behaviour. As predicted, the use of standardized materials is a very widespread solution to this problem, not only through the use of self-secreted materials, where complete standardization is possible without employing any behaviour, but also through the collection and manufacture of standard materials. In some instances, templates have been used to simplify behaviour, although getting started may be a point in the construction process where some behavioural flexibility is unavoidable. I say 'may be' because there certainly is a lack of detailed descriptions of construction behaviour in all the stages from start to finish. Information on behaviour repertoire sizes and on learning ability, both necessary if we are to compare the difficulty of different types of building, are still too fragmentary. There is much work to be done.

Now let's consider in a bit more detail, the predictions I made about building anatomy. The first of these is that the degree of adaptation shown by anatomy used for building will reflect the degree to which it is committed to building.

Birds use their beaks for nest building. Nevertheless, we saw earlier in this chapter that beaks are overwhelmingly an adaptation to feeding, with nest building as their subsidiary activity. But feeding requires considerable manipulative skill, so the head and beak can probably make most of the movements needed for nest building as a result of adaptations for feeding. Nevertheless it is a surprise to see how little evidence there is of beak specialization for nest building. The beaks of the two groups of weavers, the bird species with probably the most skilled nest building behaviour, do not show convergence in their design. The short beaks of the Old World weavers are specialized for seed eating; the straight, sharp beaks of the Montezuma oropendola are adapted to a diet mainly of fruit. If that seems a bit glib, then why do we see no obvious differences in the beaks between the sexes either in the village weaver, where the male does virtually all the building, or in the Montezuma oropendola, where it is the female that does the building and the male none? I say 'no obvious difference', but someone should actually look for small differences between the sexes in the beaks of these species. Perhaps they are actually there. Even so, the point here is that the form of bird beaks overwhelmingly indicates diet because that is how they are mainly used.

What evidence is there of specialized anatomy for skilled building in animals other than birds? An example I like is the thread-stitching anatomy of the snapping shrimp *Alpheus pachychirus*, which can stitch together mats of filamentous algae by thrusting through an algal thread held in the claws of the second pair of what, for simplicity, I am going to call 'legs', which was indeed their ancestral function. The thread, once pushed through the algal mat, can then apparently be grasped again and pulled back through. This sounds like an activity that requires care and skill and might well necessitate specialized behaviour, although we have little information on that. It also sounds as if it would require specialized anatomy, and that the shrimp certainly has.

The snapping shrimp looks like a diminutive lobster. The first pair of what I have called 'legs' are in fact a pair of massive clasping claws, one much bigger than the other for reasons that I will come to later. Behind these are four pairs of thin and delicate appendages. The last three pairs of these all look the same. They are composed of seven, hinged segments, the last of which is a simple hooked claw, and they are used simply for walking. The pair of appendages in front of these and behind the big claws, is the pair used for building. This pair differs from the walking legs firstly in having the last two segments modified to form a diminutive, finger-and-thumb, clasping claw. It also differs from the three pairs of walking legs in being longer and more flexible. What is, in the walking legs, the third segment back from the tip, has now become five segments, all jointed. Is this not good evidence of specialized anatomy for skilled building? Well, no, actually. There are a number of species of *Alpheus* snapping shrimps, most of which live not in algal mats but burrows that they dig in the sand. But they too have this highly specialized pair of limbs behind the major claws and in front of the walking legs. With them, these snapping shrimps can reach any part of the body to pick off pieces of accumulating debris. They are grooming claws.[5]

The main job of legs is to move bodies around, but in that way they may be heavily involved in building without being directly involved in manipulation. Eastern tent caterpillars (*Malacosoma americanum*) live colonially in a silken tent up in a tree. They make the silk fabric of the wall collectively and reinforce it by walking across it in a specific way, extruding the silk from their mouthparts. The path of the thread therefore matches exactly the movement of a caterpillar's head. As the caterpillar walks it naturally carries its head forwards, but at frequent intervals the head is swung in a wide arc back towards the tail end of the body, then back again. The caterpillar continues to walk forwards making head sweeps, sometimes to the right, and sometimes to the left. From time to time it changes its walking direction to the right or the left. The combined effect of many caterpillars walking over the surface of the tent in this way, trailing silk from their mouths, is the build-up of layers of silk that strengthen the tent wall.

These leg movements and head swings made by the eastern tent caterpillar look much the same as the movement of any caterpillar looking for food, except that in tent building the behaviour is more stereotyped. The point is that the legs are used to do much the same job whether the larva is foraging for food or tent building—to walk. This argument could equally be used for web building by an orb-web spider. The arrangement of threads in the completed web describes the path travelled by the spider propelled by its legs. Across the animal kingdom, legs, as organs of locomotion, have repeatedly been adapted for carefully controlled movement. This is what they did in the non-building ancestors of caterpillars and what they continue to do now, even when the caterpillar, spider or whatever is not building. No special modification may be necessary to have effective builder's legs. We have already seen with the bird example that much the same can be said about mouths. Birds have beaks that in feeding are adapted for careful manipulation; in becoming adapted for nest building as well, little modification may have been necessary.

My second prediction on the anatomy of builders was that there would be only a limited number of good solutions. The evolutionary origins of building anatomy have overwhelmingly been from organs of feeding and of locomotion, organs which generally retain their original function alongside that of building. Choose an animal builder and this conclusion is a very good fit. It is dangerous to claim any absolute rule in biology, but I can find no example of an animal builder that now uses what was ancestrally a mouth, for nothing other than building.

The use of organs of feeding and locomotion as almost the only source of building anatomy also works well across the animal kingdom. Males of the three-spined stickleback fish (*Gasterosteus aculeatus*) build nests in which females deposit their eggs. The male then tends the eggs in the nest and guards the newly hatched fry. The nest is made of plant material stuck together with secretion from the kidneys. A breeding male swims back and forth bringing pieces of material to the nest site and positioning them with its mouth; he then presses his belly against the nest material, secreting the sticky material that binds the pieces together. These various body movements (fast, slow, forwards, backwards, turning) are achieved by beating fins. That's what fins do; what mouths do is grasp objects and hold them. Fish builders do not appear to need special building fins or building mouths. The social wasps that build their nests out of paper collect and prepare the woodpulp with chewing movements of their jaws, the same jaws that they use to chew insect prey. Invariably when mouth and jaws are used for building, they retain their primary function as feeding organs. The consequence is a remarkable lack of any obvious specialization of these as building organs.

But, hold on a moment. Building anatomy cannot be quite so easily dismissed as lacking obvious specialization. Remember, there was a third prediction: that anatomical specialization will be shown where power is needed, and this will be most evident in burrowers.

Burrowing through the ground is difficult. It has been estimated that the energetic cost for an animal to burrow compared with walking the same distance above ground is between 360 and 3,400 times greater, depending upon the type of soil. Digging is hard work, and it needs special equipment to make the work efficient. We should be able to see that.

Here's another quiz question. Which of the six organisms illustrated in Figure 3.5, the three insects and the three mammals, are burrow diggers? If you say **(a)** and **(b)**, **(e)** and **(f)** then you are right as far as you go, but wrong in not also saying **(c)** and **(d)**. In fact they are all burrow diggers. Here is a supplementary question. Which of the six species live mainly underground, with only rare visits to the surface? This time the answer is indeed **(a)**, **(b)**, **(e)** and **(f)**. This degree of modification of anatomy parallels that predicted for behaviour. The degree of specialization to building reflects the amount of time spent at it.

Mammal **(c)** is a rabbit (*Oryctolagus cuniculus*). It certainly digs a burrow, but it feeds above ground and needs to be able to run fast back to its burrow if a predator suddenly appears. Fast running needs long, light legs. Mammal **(a)** is a European mole (*Talpa europea*). Its food is earthworms and it burrows through the soil in search of them. Mammal **(b)** is a Cape mole-rat (*Georychus capensis*), a subterranean-dwelling rodent that digs burrows in search of the bulbs and tubers on which it feeds.

The insect burrows follow a similar pattern. Insect **(f)** is *Cerceris arenaria*, a solitary wasp of the family Sphecidae. It digs into sandy soil to make chambers that it fills with small beetles to feed its larvae. To find and capture prey it needs to both fly and run around over and under leaves. Digging occupies only a small proportion of its time. The mole cricket (*Gryllotalpa gryllotalpa*) **(e)** and the nymphal (immature) stage of the mayfly *Pentagenia vittigera* **(f)** spend their time more or less permanently in burrows and are specialist diggers.

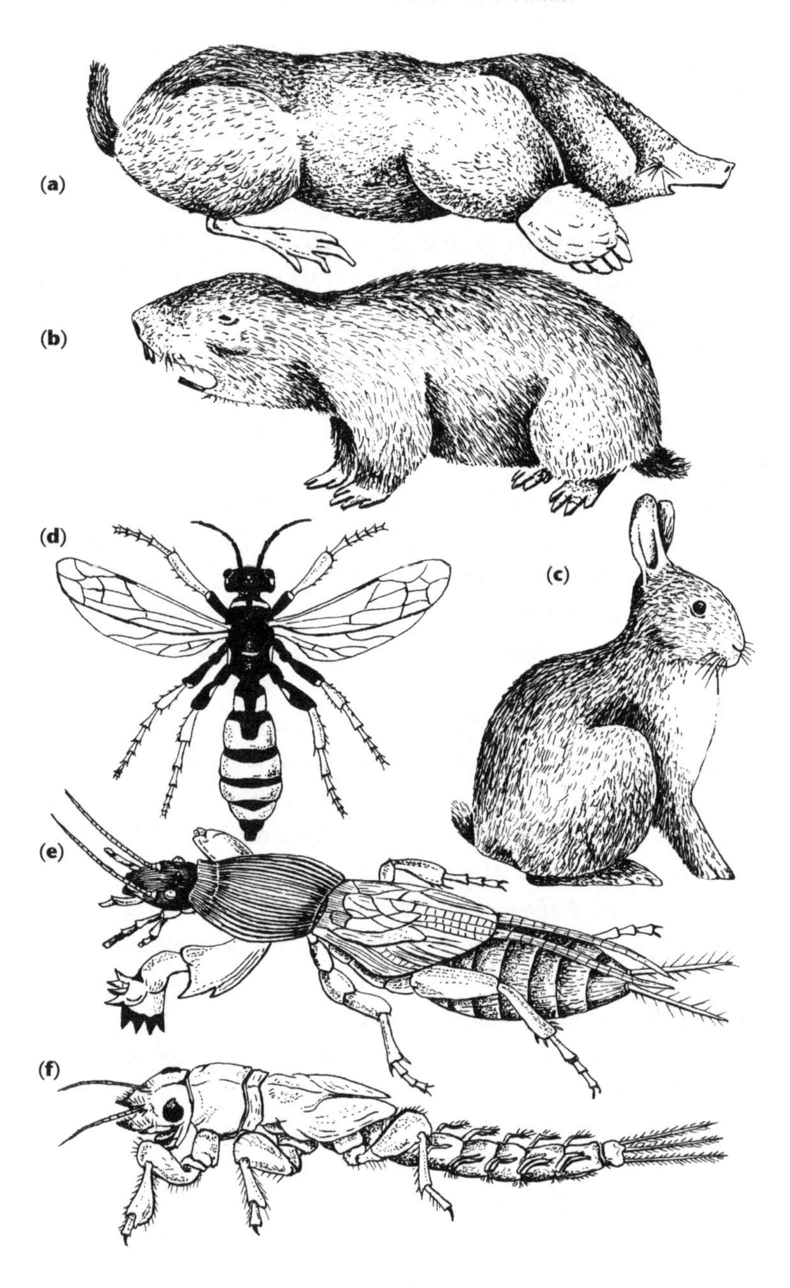
(a)
(b)
(d)
(c)
(e)
(f)

The mammals have come up with two distinct and specialized digging techniques. One involves specialized digging with the legs, the other, digging with the teeth. The mole uses front feet modified as shovels driven by very muscular, short arms. We can tell they are muscular by looking at the skeleton alone. Strong muscles must operate a strong lever; the result is enlarged surfaces on the limb bones for muscle attachment. The work done by the muscle must then be converted into a strong force at the digging end of the limb. The work delivered at the end of a lever is equal to the force times the distance from the tip of the lever to its hinge point (fulcrum). The consequence of this is that, for the same amount of work, the tip of a short lever produces a powerful force, although only moving a short distance, while the tip of a long lever moves a much greater distance but exerts proportionally less force. Moles gain more from short digging legs, rabbits from longer running legs.

Mole-rats are rodents, the defining character of which is a pair of prominent incisor teeth at the front of the upper and lower jaws (Latin: *rodere* = to gnaw; *dens* = tooth). Rodents are very largely vegetarian, using their teeth for gnawing. But in the specialist, subterranean rodents, these teeth are modified for digging. The most obvious modification has been their elongation to enable them to excavate a bigger scoop of earth with each bite, but the lips also have been modified so that they can close behind the incisors. A mole-rat can dig with its teeth without getting a mouthful of earth. Adaptation of the jaw muscles and skull for enhanced power there almost certainly is, but it is less obvious.

The specialist insect diggers show adaptations that parallel those of the mammals. There are modifications to the jaws and/or to the legs. The mole-cricket is so called because of its striking resemblance

←

Figure 3.5. Which of these six species are burrow diggers? (a) European mole, (b) Cape mole-rat, (c) rabbit, (d) solitary wasp (*Cerceris*), (e) mole cricket, (f) mayfly nymph (*Pentagenia*). (For explanation see text.)

to a mole. Because it wears its skeleton on the outside, we can see the short, stocky front limbs ending with a toothed shovel. The mayfly nymph shows a combination of modifications of jaws and front limbs. It burrows in compacted clay, using upward sweeps of its short, stout tusks as the main instrument to dislodge the clay. The broad, hairy front legs sweep the debris back along the tunnel, while the stout spur towards the tip of the leg simultaneously widens the tunnel sides.

The evolutionary origins of the digging equipment of the digging specialists described here are plain to see but, more interestingly, each represents a modification of the same parts of the body as those recruited through natural selection for skilled manipulation: i.e. walking limbs and jaws. Other specialized burrowers exist: species that dig into very hard materials such as wood or even stone. Many insect larvae feed on wood, a material so difficult to digest that growing up can take years. The larvae of a number of insect species are wood borers, larvae of moths and beetles in particular, but they universally use massive jaws for excavation and simultaneously feeding, confirming the general pattern we have already established. However, it is to the snapping shrimps I wish to return for a final example of adaptation of anatomy for power, the power in this case to drill into rock.

Snapping shrimp species typically live in natural cavities or burrows in the sand. *Alpheus saxidomus* is a snapping shrimp that lives in cavities in the rock and its home might be considered unremarkable. However, snapping shrimps are so called because one of the two major claws is modified to have a trigger action that releases the power of a huge, already contracted muscle, causing the claw to close at such a speed that it produces a shock wave through the water. This is capable of stunning or even killing passing prey such as small fish. However, the tip of this massive snapping claw in *Alpheus saxidomus* is worn and scratched by, it is suspected, abrasion against the rock. It seems that the species blasts a cavity in the rock by repeatedly holding its claw to the rock surface and pulling the trigger.[6]

Although apparently unique, this example again reaffirms themes that have emerged through the chapter: simple, repetitive building behaviour, the most evident building anatomy being that adapted for power, and the origins of building behaviour and anatomy being either from jaw movements or, as in this case, limb movements adapted previously for some other function, in this case feeding.

So, as far as most animals are concerned, you don't need brains to be a builder or much by way of special anatomy either. In that case, how can a workforce of tens of thousands of such creatures cooperate to build, as termites do, a city-state in a single building? That is the subject of the next chapter.

4

Who's in Charge Round Here?

The Petronas Twin Towers, when erected in Kuala Lumpur in 1998, were—at 452m (1483ft) and eighty-eight stories high—the tallest buildings in the world, and remained so until 2003. The matching towers are shaped a bit like two giant telescopes looking straight into the ground, their diminishing segments ascending skywards. They look futuristic in a slightly old-fashioned way, the slender horizontal skybridge that links them at levels 41 and 42 somehow reinforcing this impression. In spite of their undoubted elegance, they would be comfortable in the cityscape of Fritz Lang's visionary 1927 film *Metropolis*. Lang imagined a strictly hierarchical society with a literal and social architect as its head, below him an elite class of 'thinkers' and, at the bottom, a class of mechanics and labourers. It is a society with power concentrated at the top and commands flowing downwards. Are there any parallels with this in non-human societies? Among these, the social insects have far and away the largest workforces and this chapter is about how they are organized to build their nests.

A picture appeared in *New Scientist* magazine maybe seven or eight years ago showing a mound of the Australian termite species *Amitermes laurensis*, suggesting that at 6.7m it might be the tallest termite mound in the world. I'm not sure whether that is true or not, but the mounds of this *Amitermes* species and of African *Macrotermes* species have both been recorded at heights of between 6 and 7m. Let's think of what that equates to in our own buildings. The height of the Petronas Twin Towers at near enough 1483ft is equivalent to the height of 247 six-foot adults standing on each others' heads. The twin towers considered together are designed to accommodate something in the region of 20,000 office and other workers. It's a bit difficult to give more than a rough figure of how many termites standing on each others heads would reach to the height of 6.7m, but my very rough calculation makes it about 800; that is more than three times the relative height of the Petronas Towers. The number of termites living in such a large mound is in the region of—again figures are a bit rough—5 million; that is 250 times the number accommodated by the Petronas Towers. So, even allowing for a significant over-estimate in my termite figures, we humans are still building structures that are nowhere near comparable in relative scale to termite mounds.

The Petronas Twin Towers were conceived by a design team from architects Cesar Pelli Associates. They, with their creative imagination and reservoir of design knowledge, are the top of the hierarchy of thinkers; at the other end of the chain of command are mixers of concrete and carriers of bricks. Between these levels are a myriad of specialists involved in varying degrees of thinking and doing. There are all-important quantity surveyors that oversee cost control; there are structural engineers, and building services engineers of various kinds concerned with the design of the electrical supply, the heating, lighting and ventilation. Then there are all the skilled tradesmen and fitters at a level higher in esteem and reward than the unskilled labourers. Communication across all these specialist groups is also essential; to facilitate this, there are specialist communicators and managers to liaise, and to arrange and chair meetings.

The organization of a termite workforce is utterly different, as we shall see later in this chapter, but crucially, there is not an architect in sight yet what the termites create is truly architecture.

Consider the structure contained within a mound of the African termite *Macrotermes bellicosus*. There is the royal apartment, and there are also nurseries and fungus gardens, all enclosed in a substantial wall protecting the termites from the climatic and predatory hazards of the world outside. But the wall also cuts the termites off from light and air. The absence of light does not worry termites; their other senses, particularly touch, taste and smell, compensate for its lack. The supply of fresh air on the other hand is a serious problem, which is addressed in the architecture by the inclusion of a ventilation system.

The ventilation systems of large *Macrotermes* mounds are truly *systems,* massive in comparison to the individual insects and integrated into the mound structure. There are enormous channels and spaces permeating the mound, bringing oxygen to its heart and carrying carbon dioxide away. The multitude of chambers of the living area linked by apertures and short corridors can be regarded as the capillaries of the circulation system, where the oxygen is delivered to the tissues of the termites and fungi and carbon dioxide carried away, and the power that drives the air through the ventilation system comes from one of two possible sources. One is pressure differences within the mound; the other is temperature differences. Some termite mounds have one, and some the other.[1]

The pressure difference system is like that already described in Chapter 1 for leafcutter ants. The termite *Macrotermes subhyalinus* builds a dome-shaped mound of 1–2m high over the surface of which are several large apertures, some nearer the top, but others at the edge of the mound and nearer the base. As wind passes over the mound it has the effect of reducing the air pressure over the top of the mound relative to the base of the mound. This induces air to enter the lower apertures where it is conducted via wide channels into the heart of the mound below ground level. There it is dispersed through the chambers of the termite living space and fungus gardens. It is then drawn into larger channels again that rise towards the top of the mound, discharging the stale air.

There's a variant of this induced flow system in the mounds of *Macrotermes jeanneli.* Here there is a single exhaust channel that projects vertically 3 or 4m above the living space of the mound as a huge chimney. I say huge, because its cross-sectional area would accommodate hundreds of termites side by side. In this system, air is drawn into the base of the mound through a myriad of tiny pores, and out through the top of the chimney at an estimated rate of up to three or four litres per minute.

These induced flow systems of ventilation driven by pressure differences might be termed 'open', since the ventilation channels are obviously open where the air flows in and where it flows out. This is in contrast to the alternative ventilation system shown by *Macrotermes bellicosus*, which has no immediately obvious openings to the exterior at all. Here the air circulates within the mound according to the principle that hot air rises. Of course, if the air circulation system within the mound was entirely closed off from the outside, then there would be no way of replacing the stale air, but this ventilation system includes a multiplicity of fine channels that run vertically within the outer wall of the mound and it is there that, due to the porosity of the wall material, carbon dioxide is able to seep out of the mound and oxygen enter.

The temperature differences that drive this 'closed' ventilation system can arise in two ways, as we now understand from the savannah mounds of *Macrotermes bellicosus*. During the day these mounds are exposed to the full rays of the sun. This heats up the surface of a mound, causing the air inside the channels running through its walls to rise. This heated air discharges into an enormous enclosed space that lies above the living area. The relatively cooler air in this space is therefore displaced down through the colony chambers and fungus gardens into an enormous basement that lies under it, below ground level, a basement so big that in some mounds a grown man could fit inside.

A basement like this is apparently not simply a quarry created by a multitude of termites gathering soil to build the mound that now towers above it, but is an integral component of the ventilation system. It has the effect of cooling the air and may as a consequence have some role in regulating humidity. In any event, the circulation of the air is completed as this cool air is drawn up into the channels inside the outer mound wall behind the rising hot air.[2]

At night, it appears, the air circulation may operate in the reverse direction. As the temperature outside the mound drops, cooling the surface of the mound, it is the air at the heart of the mound that is warmer than anywhere else. It rises up through the living area and into the space above it. This drives air from the top of the mound down through the superficial channels in the mound wall, and into the basement. From there the air is drawn up into the living space of the mound to complete the cycle.

What I have described is a fully mature mound of *Macrotermes*, but it did not start out like that. The mound stands on the site where sometime in the past a newly mated young queen, together with her newfound male consort, landed after a short nuptial flight, shed their wings and dug a single small chamber in the ground. The queen laid eggs that hatched into sterile worker termites; the colony grew in numbers and the mound grew in size. Over the years the mound enlarged to reach its towering height of 5 or 6m.

Compare this with the way that Petronas Twin Towers was built. It was conceived as a prestige project to house the state-owned oil company Petroliam Nasional Berhad (Petronas). This set in motion a complex chain of decisions and actions. Contracts were signed, initiating the recruitment of all the various specialists. The building was begun, a host of differing skills were brought together, and the twin towers rose to dominate the Kuala Lumpur skyline. This culminated in a spectacular opening ceremony on Malaysian Independence Day, 31 August 1999. Only then did the building workers formally move out and the office workers move in. In the building of a 6m high *Macrotermes* mound, this transition never occurs; the building workers are the residents, and they can be said to have 'moved in' as soon as the royal pair build their first chamber. From then onwards the mound continues its gradual growth and continues to be a building site, some termites always engaged in mound construction and maintenance.

Consider two possible extremes of organization of the termite workforce. In the first, every individual can do any building task and is genetically equipped to understand where and how to do it. This should simplify the system of coordination between different parts of the workforce, since any individual could judge for itself what was to be built. But this seems to imply that each insect would have some conception of the overall plan of what it is involved in building—that the individual carries with it the plan of Metropolis. Given the limited learning capacity of a termite, that would necessitate the inheritance of a great deal of complex information and require each worker to have a large brain to instruct its building behaviour and monitor the consequences.

In the second workforce model, groups of specialists are assigned each to the building of a particular sort of architectural feature so that none needs to know the whole of what is to be built. But this raises a different problem, coordination between the specialist groups to make sure that work is carried out in the right order in the right place. Seeing the enormously elaborate communication networks needed to direct a large human civil engineering project, this appears unlikely. It would also probably require large and specialized brains. Is there in fact any evidence that the brains of social insects are larger than those of their nearest solitary-living relatives or that their repertoire of behaviour is any more complex?

The answer is no. The nearest living relatives of termites are actually the cockroaches; for the social-living ants, bees and wasps, it is solitary bees and wasps (all ant species found today live socially although they vary greatly in their colony sizes). Some cockroaches live in family groups or in large aggregations that have a degree of social coordination, some dig burrows, although most build no structures at all. Some cockroaches are very big—6 to 8cm long, others are nearer the 6–8mm of an average termite, but there is no evidence that the brains of the mound building termites are especially large in relation to their body size compared with cockroaches. The brains of honeybees are, in fact, rather large in comparison with those of similar sized beetles, but are much the same size as those of solitary bee species, that build much simpler nests. However, it seems that honeybee brains are particularly developed to be able to learn and remember the location and quality of food sources, rather than to deal with the problems of cooperating with hive mates to build honeycomb. Large colonies of social insects bustling with activity do give a strong impression of great behavioural complexity, but this appears not to be the case. A study on ants comparing the number of different behaviours in species with small and with large colonies found that, although there was evidence of greater behavioural diversity in the latter, the difference was slight. Social insects are able to build their enormous and elaborate nests with brains not obviously different from solitary insects, many of which build nothing.

This is not to say that social insects are unable to learn and remember aspects of the nest structure. A swarm of wild honeybees when looking for a new nest cavity can, for example, calculate the size of a cavity in a tree. When a colony acquires a new queen, it splits, half departing with the old queen to a new cavity that has been located by scout workers. The scouts obtain an estimate of the size of the cavity by walking round the interior walls. This we know from an ingenious laboratory experiment which allowed bees to explore the inside of a cylinder with a rotating wall. When a scout bee walked in the direction of rotation of the drum, it got back to the entrance with little effort and therefore underestimated the size of the cavity (the equivalent of walking up an 'up' escalator rather than the stairs). When a scout walked in the direction contrary to the rotation of the drum (like walking up the 'down' escalator), the extra work it had to do to get back to the entrance caused it to overestimate the size of the cavity. We can see the estimate they make because when a scout returns to the home nest it advertises a cavity of acceptable size by performing a dance on the comb surface in the same way it would do when reporting the discovery of a good food source. In the wild, bees that have discovered alternative suitable sites will also advertise. Recruits will then visit these sites and return to advertise their preferred location. Eventually, one site receives sufficient support judged by the dances of returning scouts that the swarm decides to occupy it.

This does show at least some ability by insects, albeit quite limited, to learn and retain information about spatial relationships. But can a termite remember its way round even part of its home mound? We really don't know. As you can imagine it would be virtually impossible to track an individual termite in its home mound and very difficult to give it maze learning tasks, isolated from its nestmates. Clearly termites are very good at *finding* their way in the home mound but this is not the same as *knowing* their way around. You know your way around your own home, but you have in your mind enough knowledge of houses to quickly find the kitchen in the house of a total stranger. A termite might be genetically equipped to understand similar architectural signposts. We should never underestimate the capacity of animals simpler than ourselves to come up with effective yet simple alternatives to our own. This is nicely illustrated by the way that ants of the species *Temnothorax albipennis* estimate the size of a potential new nest site.

This is a minute ant of body length about 3mm that lives in colonies of no more than 500 or so in cracks in rocks. These are literally 'cracks'—the nest cavities are almost two-dimensional, giving little more than headroom for the ants themselves. This is convenient if we want to answer the question of how the ants tell that a cavity is big enough to house a colony. If we provide an experimental nest site in the laboratory with the correct headroom, we can simply ask the ants if the floor area is big enough.

If a colony of *Temnothorax* is looking for a new nest site they, like the bees, send out scouts that inspect possible cavities. On finding a suitable one, a scout returns home to recruit new inspectors, but how do the ants judge that a cavity is suitable? An ingenious programme of experiments tested three rival hypotheses of how an ant might work out the surface area of the cavity floor. The experiments were designed to test which hypothesis could explain the results and which ones could not. The three hypotheses were . . . Well, before I tell you, just take a moment to think of one way in which, on entering a room in complete darkness, you could roughly estimate how big it was.

The three ways the experimenters speculated a scout ant could do it were:[3]

Firstly: That the ant walks round the wall till it gets back to where it started, so measuring the *length of the internal perimeter*. You could do this by taking one shoe off and feeling your way round the room till you discover your shoe again.

Secondly: That the ant walks directly away from the wall until it reaches another wall. Doing this several times from different points gives a *mean path length* for the separation of the walls and from that an estimate of cavity size.

Thirdly: That the ant employs the principle of *Buffon's needle* proposed by the eighteenth-century French naturalist and mathematician Georges-Louis Leclerc, Comte de Buffon. (Comte de Buffon demonstrated that if a given area on a page was marked out with a series of parallel lines then, when a needle of known length was thrown repeatedly on to the page, the given area could be calculated from the frequency with which the needle landed on a line. The ants, it was reasoned, could use this principle to calculate the size of a potential nest cavity if they could lay out a pathway of known length over the area on an initial visit then detect it on a second exploratory visit.)

The first hypothesis was rejected because the ants were able to distinguish between a circular arena and one of the same perimeter but with the walls pushed together to enclose a long narrow space of much smaller area. The second hypothesis was also rejected because the ants found equally acceptable arenas that were of equal area, even if one had a straight barrier placed part way across the middle of it. If the ants were using the mean path length estimate, they would frequently hit the extra barrier and so lower their estimate of the arena size.

The Buffon's needle hypothesis was, however, supported. Furthermore the experimenters were also able to show how the ants do it. On its first inspection of the potential nest chamber, a scout lays down a scent trail of a quite standard length that criss-crosses the area. This inspection provides the set of lines. A scout will, however, re-inspect the chamber, laying down a second trail that allows it to estimate the cavity area from the frequency with which its first and second sets of scent trails intersect. To do this not only requires a little piece of 'computation' by the scout but also, in case other visiting ants are also laying trails, for the scout to be able to recognize its own individual scent trail. This was also shown to be the case.

The Comte de Buffon was incidentally one of the scientific giants of the eighteenth century. He is best known for his monumental encyclopaedia of the living world, *Histoire naturelle, général et particulière*, published in thirty-six volumes over a number of years, starting in 1749, but in his writings he also suggested that the world might be a lot older than the 6,000 years that Christianity then held, and that humans and apes might possibly share a common ancestry.

Let's get back to how a group of social insects might collectively build a large nest. What we have so far established is that individual insects might be able to obtain some appreciation of the size of the space in which they are working. What we now need to establish is whether, and if so how, a group of individual insects can operate as a workforce. A pre-eminent feature we identified in the human workforce was communication, and biologists have a great deal of evidence of animal communication. One of the recipients of the Nobel Prize for Medicine and Physiology in 1973 was Karl von Frisch, for his work on the 'dance language' that honeybees use to inform nestmates of the distance and quality of food sources, or indeed new potential nest sites, through the orientation of the body and frequency of abdomen vibration of a dance carried out inside the nest after returning from a foraging trip.[4] Ants are known to use a variety of volatile organic molecules as chemical signals (*pheromones*) to coordinate colony activities such as foraging for food and defence of the nest. Termites use pheromones, as well as touch and vibration signals in communication. We should expect that such signals will similarly be used to coordinate building. However, in this context there is potentially another source of coordinating signals, the nest itself.

Suppose a wasp detects a hole in the nest envelope. It collects a load of nest material and with it partially repairs the damage. The nest now looks different, but needs some more material in a slightly different place. The wasp detects this, collecting and adding more material, bringing the repair a little nearer to completion. The wasp, through a chain of stimulus and response, is in a dialogue with the nest. In fact, it does not need to be the same wasp that collects and applies each load. A group of wasps could be involved, each collecting and repairing the damage. The nest can coordinate the workforce without its occupants needing to communicate with each other. Organization of this kind, where several individuals can independently but in parallel begin and end a task, in this case nest repair, is called *parallel-series*. The alternative to this method of organization is called *series-parallel*, which is where one group of specialists start the behaviour sequence and then pass it on to another specialist, and so on. In the repair of the wasp nest for example, there might be just two stages, with a group of wasps collecting material, which they then pass on to a group that use it to repair the nest. There is a theoretical advantage to this over parallel-series. In series-parallel, the failure of a single specialist material collector to complete its task only interrupts a small part of the building effort, because each builder can get material from any collector. Where an individual carries out the whole task from beginning to end, failure at any stage is a loss of that whole sequence. There is an additional potential advantage in having specialists. They can become masters of their trade, making each stage of the sequence more efficient.

Social insects do in fact show job specialization, achieving it in two ways, by body shape and by age. A good example of the former is illustrated by the body sizes of workers in the leafcutter ant *Atta sexdens*. The largest of these has a head width of greater than 3.0mm, ten times that of the smallest workers. Your head has a side to side width of about 20cm. Consider standing next to someone with a head diameter of 200cm; that is probably wider than you are tall.

Although in the leafcutter ant colony, workers exhibit the whole range of intermediate sizes, there are four typical head widths: 1.0, 1.4, 2.2 and 3.0mm. Between them they divide up the colony tasks. The smallest class care for the eggs being laid by the queen, feed young larvae and tend the fungus gardens. The 1.4mm head-width group are described as *within-nest generalists*, and perform some nest building activities; the next size class are *foragers and excavators*, which obviously engage in some tunnel-digging work, and the largest class, with their massive heads and jaws, are soldiers, defending the colony against predators.

Having a body specialized for the job probably seems like an excellent method of creating specialists. Their body is their toolkit; each specialist has its own. However, specialization by body shape is much less common in social insects than is specialization by age. In fact, less than 20 per cent of ant genera show any marked degree of specialization by body form (morphology). In termites, too, morphological castes are a minority. In bees and wasps no morphological castes at all are discernible in the workforce. The queen is always recognizably larger than the rest, but the workers all look the same. The lack of morphological specialization in bees and wasps may be something to do with them needing to fly. This is a constraint on extreme body shapes that does not apply to the earthbound ants and termites. Whatever the reason, both bees and wasps only show evidence of job specialization by age.

The immature stages of bees, wasps and ants are, as helpless grubs, looked after by adults, so it is the changing work of adults with age that we are looking at. In honeybees for instance, a worker spends the first few days after emergence from its cell, cleaning empty comb cells from which fellow adults have also recently emerged; she also feeds the grubs in the occupied cells. Over a period of three to four days after emergence, she then develops glands on the underside of her abdomen for producing wax, allowing her to move on to being a builder, combing the scales of wax from her belly and transferring them to her mouth. There they are chewed and mixed with a salivary secretion to create comb wax which is then used to build new cells. But the wax glands soon begin to shrink and, beyond fifteen days of age, a worker honeybee is spending most of her time on activities outside the nest, foraging for pollen to feed the larvae, and for nectar which is regurgitated and stored in the wax combs as honey.

A similar pattern of age-related job specialization can be found in wasps. The wasp *Ropalidia marginata* is a primitively social wasp where 'primitive' in this context means showing the ancestral condition before a true sterile worker caste evolved, with colony sizes reaching perhaps forty adult females. In such colonies, all emerging females are potential egg-layers on the nest or founders of new colonies. Even so, age-related job specialization is apparent and follows much the same pattern as seen in honeybees. A newly emerged female tends to be a feeder of larvae, taking food from incoming foragers and distributing it among the cells. Later, she will probably go through a phase of building new cells with material brought in by paperpulp foragers. Then these nest builders begin to leave the nest themselves to collect building material, finally moving on to foraging for insect prey, the characteristic food of wasps.

You may now be convinced that job specialization by age in social insect colonies is an effective and simple method of ensuring that there are individuals dedicated to each of the nest jobs, care of young, nest building and foraging. But suppose a colony of honeybees has been in a phase of rapid expansion with a lot of grubs to care for but a very young group of workers. That would create a temporary excess of cell cleaners and cell builders when the priority is really for foragers. In such a situation it would be more efficient if the job undertaken by a honeybee worker was not strictly determined by its age, but could take into account the current colony needs. To test whether there is this flexibility in the honeybee colony workforce, the populations of different age classes in a colony were experimentally altered to see if the bees then shifted their activities. They did.

In one experiment, a group of eight to thirteen day workers were added to a colony—that is, of workers normally engaged on within-nest activities. The result was that the older of the nest-bound bees of the original workforce immediately shifted to out-of-nest activities, while in unaltered control nests they waited until they were older. The converse experiment produced an equivalent response. When younger (nest-bound) bees were removed from the workforce of experimental nests, the younger of the experienced foragers reverted to within-nest activities. It seems that the activities of honeybee workers are related to age, but not strictly determined by absolute age. They tend to do more out-of-colony work with increasing age, but choose jobs typical of their age according to the current needs of the colony.

So far we have looked at all colony activities, but what about the organization of nest building? In social wasps the building material is principally wood pulp, although salivary secretions may also be incorporated. This pulp needs to be collected by foragers. This is more complicated than the situation in honeybees where the nest material is self-secreted and one bee can combine wax secretion, wax processing and comb building. So in wasps, nest building requires a combination of on-nest and off-nest activities, and that involves some collaboration between different groups of wasp, each specializing in some task in the nest building process. This may differ in detail between wasp species but, thanks to the distinguished work of Bob Jeanne from the University of Wisconsin, Madison, USA, and co-workers over a number of years, we do have very good experimental evidence of the organization of nest building in the tropical New World wasp *Polybia occidentalis*.

Colonies of this wasp can reach a size of two or three hundred, and they form new colonies by swarming, in the same manner as honeybees. Soon after swarming to a new nest site, the wasps start to construct a nest. When completed, it will consist of a disc-shaped comb enclosed in an envelope, with a single entrance at the bottom. As the colony grows, new combs will be added in a stack below the first. The workforce of *Polybia occidentalis* shows the age-related trends in job specialization that we have already seen. Nest building is a job that is generally undertaken by wasps of up to twenty days of age. Beyond this, outside-of-nest activities predominate, but there is more to it than that.

What might be called the 'trade' of nest building turns out to consist of three 'tasks': actual nest building, foraging for wood pulp and collection of water; the role of the water being to adjust the consistency of the pulp. These tasks, it turns out from the observation of wasps individually marked with coloured paint spots, are done by different groups of wasps.[5] It is a clear example of job specialization employing the series-parallel principle. Pulp collectors and water collectors hand on their materials to builders who complete the nest construction sequence. The theoretical merits of the series-parallel over parallel-series organization of sequential tasks has already been mentioned, that of greater likelihood of job completion and higher quality of work done by specialists, but here another benefit is evident: efficiency in pulp collection. It turns out that a worker wasp can collect a ball of wood pulp that is 2.6 times the size that a worker can handle when adding material to the nest. The pulp forager therefore, on arrival at the nest, divides its load between builders, a much more effective method than for parallel-series organization where each builder would be bringing to the nest only as much material as it alone could add to the nest, a much smaller load than the maximum it could carry.

The task groups need now to integrate effectively with one another. There must be an appropriate number of builders to accept the pulp loads that those foragers bring. At the same time, there needs to be a matching number of water foragers to supply the builders with the water needed to soften the pulp loads. This integration is likely to be especially difficult in small colonies because, in comparison with a large colony, the chances of a builder meeting in quick succession both a pulp and a water forager are low. This will result in longer waiting times for all three task specialists in small colonies, unless individuals are able to be flexible and switch tasks. This turns out to be the case. In large colonies there is very little task switching, but in small colonies it is quite common—a principle that we can recognize in human business management.

This of course raises interesting questions about how the flexibility in the workforce is achieved. Well, experiments show that removing a significant proportion of any one of the three task groups in a *Polybia occidentalis* colony has little effect on the rate at which the tasks are performed. This means that the lost specialists are being replaced by other individuals that have either been engaged in some other task, such as food foraging or, contrary to our perception of social insects, individuals that have been hanging around the nest doing nothing.

The study of marked wasps has shown that wasps leaving the nest to forage are seeking one of four resources (pulp and water for nest building, pollen and nectar for food), and that when individuals switch tasks it is generally within the feeding function (pollen to nectar or vice versa) or within the nest material function (pulp-water). When the nest is experimentally damaged, so that additional recruits are required to gather the two nest building materials, they do in fact come from a pool of inactive individuals. Furthermore, those individuals are known to have previously foraged for building materials but not for food; they are specialist reservists. The life of a social insect is not necessarily one of constant industry. You may wonder at this seeming inefficiency, but what are reservists needed for? They are there in case of an emergency. In this study the nest was experimentally damaged, but these tropical wasps live in a hostile world. Natural nest damage (for example from storms), and the need for rapid nest repair, are probably not rare occurrences.

We can now see how nest building in this wasp species is organized, but the coordination of it requires communication, so how is this done? How, for example, can an inactive individual become aware that it is needed for whatever task or, when it is already occupied, that it is no longer needed? Does it get a signal from other wasps, from the nest itself, or what? To unravel this, further experiments were undertaken in which not only were the majority of each of the three task forces in turn removed, but water availability was increased by adding some water to the nest envelope close to where the wasps were repairing it, and wood pulp availability supplemented by placing a supply of it on a little platform just beside the nest.[6]

The results of these experiments showed that only members of the nest-building group obtain information directly from the nest. When they locate damage they remain there to engage in repair. Their numbers grow until this inhibits others from joining in. Experimentally providing additional pulp has the effect of reducing the numbers of pulp collectors, suggesting that this task group are dissuaded from collecting pulp when they experience low demand for their wares from builders, communication in this case being between task groups.

Providing water by spraying the nest had the effect of reducing water foraging, but so also did removal of pulp foragers. The level of activity of water carriers seems to be determined by the level of demand they experience from pulp carriers to whom they give their water, communication again being between task groups. Communication is, it transpires, very simple. The whole building sequence is initiated by a nest-building group stimulated by the discovery of nest damage; they then regulate their own numbers and, through their demand for building materials, the numbers of pulp and water foragers. The information an individual receives is uncomplicated and the decision it has to make offers few alternatives.

A picture is emerging of the organization of social insect nest building that is radically different from our own. There is no leadership. No individual or individuals can be said to be 'in charge'; there is no hierarchical structure or line management. Where there is a sequence of activities to be completed, individuals are stimulated to become active or inactive through very simple signals from other colony members or by the nest itself. The question now is how can such a workforce produce architecture?

Of the social insect nests, termite mounds are among the biggest and are arguably the most architectural, at least in the sense of having areas of the mound of quite different design that do quite different things: ventilation channels, cellar, royal chamber, nurseries and fungus gardens. How, given that the organization of termite workforces is more likely to resemble that of wasps than that of humans, can they possibly achieve this? One possible way is for every termite to be equipped with a *blueprint* (architectural and engineering plans used to be printed on blue paper). If each individual inherited the same blueprint of what was to be built, the absence of a termite architect in a colony would not be a problem. However, there are a number of difficulties with such a system. In the first place, a mature mound of termite species such as *Macrotermes bellicosus* is so large and complex that it is hard to imagine how all its architectural detail could be encoded in the genetic material passed from one generation to the next. Second, the architecture of a mature mound is only what is finally achieved after several years of growth from an initial single cell with the founding royal pair. At each stage its relative proportions change as the relative importance of the different elements shift. Could all termites really be equipped with all the different blueprints needed for the different stages of colony growth? A third major problem for a termite builder is in knowing how much of the blueprint has so far been completed. Insects may have some powers of spatial memory but could a termite possibly decide what and where to build on the basis of a prior inspection of the nest? Based solely on the time it might take a termite to do such an inspection, this seems impossible.

The question of whether animal builders had a mental blueprint of what they were attempting to build has a long history of debate in the field of animal behaviour. An English researcher, Bill Thorpe, at the

University of Cambridge in the early 1960s, was of the opinion that birds in building their nests must use a mental image, and then strive in their building behaviour to match their effort to reach the goal. An extreme opposite strategy to this would be for builders to interact with the structure solely at a local level, detecting a building stimulus that evoked a simple building response, the changed stimulus leading to a new response and so on. Such a stimulus–response chain could lead eventually to nest completion and would require no mental image, nor indeed would it necessarily require any memory of what had so far been achieved.

The test for the mental image hypothesis has always been evidence of innovation. If we experimentally damage the nest of a bird to create a structure that is unlike any stage of the normal building sequence, can the builders come up with the most economical route towards completion of the nest? If so, it is argued, then the bird must know where it is trying to get to, and indeed be able to understand what problem the experimenter has set.

The results of these experiments have always been equivocal. The Baya weaver builds a hanging nest with a downward projecting entrance tube. In an experiment conducted in the 1960s, nests were damaged by removal of the entrance tube and the antechamber above it, and a large hole was cut in the nest chamber on the other side.[7] Some birds repaired all the damage to restore a normal nest, apparently demonstrating an understanding of the overall nest design. However, some blocked up the entrance, leaving an entrance hole with no tube on the other side, while one blocked up both holes so shutting itself out.

On the other hand, evidence of building sequences being entirely composed of stimulus–response chains are not fully convincing either, but at least we know that some bits of building behaviour are guided in this way. The nest of the paper wasp *Polistes fuscatus* takes the form of a single comb of downward-directed cells suspended from a fine paper stalk attached to an overhead support. A female *Polistes*, having selected the nest site, flattens out a small pad of paper pulp on the attachment surface. To this pad she attaches a downward-projecting stalk which is further extended as a flattened tongue of material. The first cell of the comb is built on one side of this flat surface, the second cell on the other. This highly predictable sequence looks like a stimulus–response chain. After this, the pattern of further building varies between individuals. Cells may be added here or there to the edge of the comb; periodically the stalk suspending the nest is reinforced.

Some researchers question the rigidity of even this opening sequence of *Polistes* nest building. They believe that the wasp may already be checking on all aspects of progress of the structure. After this initial phase, the behaviour sequence certainly becomes less predictable. The female wanders over the comb surface from time to time adding further material, but what is she doing? Is she inspecting and comparing progress across the whole structure or, as those favouring a stimulus–response interpretation would argue, just wandering about until she finds a stimulus that evokes a building response? We are still quite unclear as to the extent to which a social insect, such as a wasp or termite, can make some kind of general assessment of building progress before adding to the structure. However, what we do know from observing the behaviour directly and from the study of virtual builders in computer simulations is that a surprising amount of the creation of nest architecture can be explained in terms of very simple local responses.

The French have a strong tradition in the study of social insects and it was a pioneer in the study of termites, Pierre-Paul Grassé, who in the 1980s demonstrated that worker termites, removed from the nest and placed in an arena with nest material, initially place drops of it at random. Gradually, however, as the termites continue to deposit new pieces or move ones already deposited, quite evenly spaced larger blobs of material begin to appear. These attract new material which is added to the tops of the blobs to create pillars. Where two pillars are neighbours, they then grow towards each other at the top to create arches. This process can be largely explained by the termites having included a pheromone in pellets of building material, which attracts other termites to deposit their load at that spot. Initially, all locations are equally attractive but this system operates by positive feedback: as soon as one point gets slightly more material it attracts more termites to it to deposit more material on it.

Grassé called this system of building *stigmergy* (roughly meaning 'focus of work').[8] This is a building principle in which an individual is seen as responding with building behaviour whenever the existing structure presents a 'stimulating configuration'. Such a building process would demand very little of the termites in terms of understanding or decision-making, it does not even require any direct communication between the members of the workforce. It is a stimulus–response account; the termites just go looking for work, getting their instructions on how to build from within themselves, and on where to build from the building itself. A job once started can be continued and completed by anyone, without the need for communication.

Another experiment conducted by Grassé showed how a different architectural feature of the termite mound might be created using the same principle—in this case the royal cell that contains the queen of a colony of *Macrotermes natalensis*. This chamber is roughly the shape of the queen but with a fairly generous space round her on all sides to allow the workers to attend her. The queen is, in comparison with the workers that scurry round her on all sides, an enormous bloated and gently pulsating, white sausage, a giant factory for producing eggs.

Grassé demonstrated that, if the queen is removed from the nest and placed in an open arena with workers, they will build a new royal chamber round her like the one from which she was removed. They begin by building a low wall, leaving as much space round her as there was in her original cell. As they continue to build up the wall, they arch it in over her body to create a domed roof that closes over her. He concluded that it was a pheromone emanating from the whole body of the queen that determined the position of the wall. This cloud of scent set up a gradient of the vapour around the female, strong close to her, diminishing with distance. A threshold value of the pheromone on this gradient determined the position of the wall. The cloud of the pheromone enveloping the female defines the size and shape of the queen cell. It is another example of a building template, similar to the caddis larva and village weaver using their own body proportions to define the boundaries of their built structures. The queen termite pheromone cloud provides the 'stimulating configuration', and the architectural outcome of a royal chamber can be explained as a group of individual builders using the principle of stigmergy. I say deliberately 'can be explained'. A computer model incorporating only worker responses, the pheromone template created by the queen, and the attraction of worker pheromone in the building material, can generate a virtual royal chamber. There may be other complexities in the organization of the normal building process, but the point of the computer model is to show that such elaborations are not essential to produce the observed outcome.

Theoretical models like this are therefore very useful tools for testing the degree of architectural complexity that can be generated using very simple rules, in particular rules that don't require members of the workforce to pay any attention to each other but just to converse with the emerging architecture. A dynamic model similar to the one that generated a virtual royal chamber for the termite queen, can also create a virtual corridor. To do this, the model assumes that the building process is occurring not in still air but in moving air. If we imagine worker termites placed in an open arena starting to build pillars and then create an arch where a light wind is moving through it, the consequent movement of the nest material pheromone encourages building on the downwind side. This tends to steer the movement of termites themselves in that direction. Termites, we know, lay trail pheromones. If that feature is incorporated into the model, it stimulates the directional movement of more termites, which place building material loads beside the trail, creating a corridor.

Let's take a reality check. This is a computer simulation not a real colony of termites. However, notice that, even when the building rules used by the virtual termites remain constant, the architecture produced varies according to environmental conditions, in this case by the movement of air and the resulting movement of the termites themselves. Communication between termites is again shown to be unnecessary.

One other conclusion is also very important. We don't need to claim that the termites had any intention of creating a corridor or that a worker termite had any concept of a corridor. The corridor is an *emergent property* of a mass of individual responses to local stimuli. If that seems surprising, then it shouldn't. Consider the hand that you are holding this book with; when you were a foetus and your hand was a simple paddle, did the cells that then comprised it 'understand' that they were going to build a hand? They were all equipped with some inherent rules and responded to some more or less local stimuli: templates, gradients and so on. They each responded by multiplying, moving, differentiating or dying, and a hand emerged.

Emergent properties, that is, appearance of a higher order of organization as a consequence of lower-order instructions, is a powerful phenomenon as can be seen not just in biological but in dynamic physical phenomena. In the permafrost regions of the arctic tundra there are places where the ground is divided by sharp, straight lines into a pattern of polygons from about 10 to up to 50m across. These look like the result of some human intervention, but in fact come about simply as a result of the repeated freezing and melting of water. In winter, extreme low temperatures cause the ground above the permafrost to crack. In summer, water from the melting surface snow and ice runs into the cracks where in winter it then forms wedges of ice, so expanding the cracks. Over hundreds of cycles, a settled regular pattern of cracks emerges, dividing the ground into polygons.

The ant *Temnothorax albipennis* provides not a computer simulation, but a physical example of how from simple, local building rules, a higher-order structure emerges. This was the ant that earlier in the chapter was found to use the principle of Buffon's needle to estimate the dimensions of a potential nest cavity. When a colony decides that a cavity is suitable and moves into it, colony members first all cluster together. Then individual ants venture out in search of the tiny stones that will make up the wall with which to surround themselves. On returning, a forager brings its particle right up to the group of its nestmates, touches them, turns through 180°, moves outwards for one or two body lengths and drops its load. As stones begin to accumulate, a forager will use one held in its jaws to bulldoze other sand grains outwards to compact the wall.

The presence of the stones also stimulates the deposition of others. Where they are dense, new stones are more likely to be deposited. Where they are scattered, a stone is more likely to be removed. The result is a more or less circular defensive wall all around the ants that reaches the ceiling of the cavity. Trail pheromones, which inhibit stones being dropped, ensure at least one path or doorway through the barrier. A computer model incorporating only these simple rules generates the same pattern of building. With the cluster of ants providing the template, the system is *self-organizing*, giving rise to the enclosed nest space as an emergent property of lower-level actions.

There is an interesting digression to this story, which concerns the selection rules for the size of stones used by *Temnothorax albipennis* to build the walls. An experiment was carried out on these ants, which offered them as building material only stones of two sizes: large ones with a grain diameter of nearly 1.0mm, and small ones with a diameter of a bit more than 0.5mm, both within the size range acceptable to the ants. It was predicted that the ants would, given a choice, greatly prefer the bigger size because both take about an equal time to collect, so more material can be collected in a given time by concentrating on large particles. It was also predicted that the ants would show a stronger preference for larger particles the greater the recovery distance. The reasoning here is the one that results in you being prepared to go to your local family grocer for a carton of milk and half a dozen eggs but, when you drive to the out-of-town supermarket, always coming back with a car load of provisions that will last the family several days. These predictions were, however, poorly supported.

The *Temnothorax* ants, when collecting stones from whatever distance, picked up both large and small ones. They did certainly collect a greater proportion of large ones with distance, but as wall building progressed the tendency to collect small stones increased. Stones had been selected to achieve something different from simply economy of effort. This, it transpires, appears to be the structural integrity of the wall.

Take a bucket of dry sand and pour it carefully on to the ground. As the cone of sand builds up, so its sides get steeper. Then, at a certain point, avalanches of sand begin to launch themselves down the sides. After that, no matter how much more sand you pour on to the peak of the cone, and no matter how big the cone gets, the steepness of its sides will not increase. You have discovered the angle of maximum stability for your supply of sand.

This angle is different for different sizes and shapes of sand grain, and for different mixtures of sand grain size. For the sand grains offered to the *Temnothorax* ants, it turns out that the angle of maximum stability is greatest for a mixture of about equal parts of large and small grains, which further investigation showed to be the mixture closest to the ratio with the densest packing of particles, i.e. where there is the smallest percentage of unit volume occupied by space between the particles. Selection of small as well as large stones by the ants is apparently determined by the need for maximum wall stability, and not simply by the need to collect building material quickly.

Builders, like this species of cavity nesting ants, are proving very convenient for studying some of the basic principles by which social insects build nests. Experiments with colonies of termites to investigate the creation of mound architecture seem far too complicated, but that after all is what we want to be able to explain. In the meantime computer models have shown just how much architecture can be generated with very unsophisticated virtual builders. This is nicely illustrated by the so-called *lattice swarm* models designed to simulate the building of nests by wasps.[9] The virtual workers in these models are endowed with very limited capabilities: they cannot communicate with one another, they have no memory and respond only to local stimuli. They are, if you like, 'stigmergic' in their building behaviour, like the real termites building their pillars, arches and royal chambers. Having been programmed with a simple set of building rules, the virtual wasps move over the surface of the virtual nest in search of a 'stimulating configuration' on which to add their own contribution. In these wasp nest models this contribution is a 'brick' or building unit resembling a single cell of the nest comb. The question is: can a colony of such creatures create anything like a wasp nest?

The answer, perhaps surprisingly, turns out to be yes. Given certain sets of instructions (algorithms), the virtual wasps produce designs that show regularity in the shape and arrangement of combs (Figure 4.1). These are so-called 'coordinated' algorithms. One design feature of the virtual nests generated by co-ordinated algorithms is their *modularity*, i.e. the repetition of nest elements in a regular arrangement. This is a very characteristic feature of nest architecture across wasp species, for example taking the form of horizontal combs stacked one below the other. These, it seems from the models, can be explained as an emergent property of simple behavioural rules. Coordinated algorithms, however, turn out to be rare; the great majority of algorithms prove to be 'uncoordinated'. Each of these produces a more irregular architecture which is different every time the programme is run.

The comb designs in real wasp nests are, as it happens, even more diverse than those simulated by the lattice swarm models, and a feature of some of them is that the combs are not flat but curved, in one or two planes. The nest design of *Agelaia areata* is a comb that

Figure 4.1. Virtual wasp nests generated by coordinated algorithms exhibit modularity.

Reproduced by permission of Guy Theraulaz

continues to curve as it grows, becoming eventually like a loosely rolled scroll of cardboard. Curved surfaces would appear to be quite complex structures to make. For example, Foster and Partners arched glass roof over the central courtyard of the British Museum, which opened in 2000, covers the space between the edges of the square and a central circular pavilion with 1,656 triangular glass panes, each uniquely shaped. How can bees and wasps create curved combs by the repeated addition of apparently uniform building blocks, the brood cells?

The cells of a honeybee comb are, famously, hexagonal. This is a way of dividing up a plane surface into regular repeated units, without having gaps between them. There are two other ways of doing this, squares of course, and triangles but, where a triangle, square and hexagon are drawn to enclose the same area, the circumference is least for the hexagon. Therefore the wall of a hexagonal cell uses less wax than one of triangular or square section with the same volume. So, hexagonal cells are economical in the use of materials and of course are a better shape than either square or triangular cells for the plump bee larvae that grow inside and eventually fill them.

If the flat comb surface is to become curved in one plane, then the surface shapes of the hexagons, squares and triangles will become curved—just roll up a piece of squared paper into a tube and you can see the effect. If the surface is to curve in two planes to form a dome, then no single regular identical unit is possible. They need to be individually adjusted. How can wasps do that?

In 1985 a remarkable molecule of pure carbon was discovered, composed of sixty carbon atoms linked together. It proved to be a spherical molecule made up of sixteen identical hexagons and twelve identical pentagons, forming a skeleton reminiscent of the geodesic dome architecture of Buckminster Fuller. It was quickly dubbed 'buckminsterfullerene' or 'bucky ball'. If you find its structure difficult to visualize, then look at modern soccer balls—one version is designed as a bucky ball, with hexagons often in white and the pentagons in a conveniently contrasting black. Only these two units are necessary. Combining some identical hexagons with some identical pentagons creates the domed surface.

We know that the curvature in at least some wasp nest combs is the result of including pentagons among the hexagons. Earlier in the evolution of the design, the pentagons could have been regarded as 'mistakes'; however, the curved combs apparently proved advantageous in the environment of these ancestors and the 'mistake-makers' persisted through the generations.

So is that all there is to the creation of curved combs, an emergent property of individual workers adding a mixture of hexagonal and pentagonal cells? We realized at the start of the chapter that social insects must build in a radically different way from ourselves. We can now at least see that there are ways that a very large workforce of small-brained creatures could build a complex structure using very simple principles of organization. Further research may, of course, reveal complexities in the way that termites create mounds or wasps build nests. However, a theme is emerging in this book of the surprising sophistication in construction that can be achieved by animals with small brains and a limited behaviour repertoire. This is a strong theme in the examination of trap making by animals in Chapter 6, and forms the background for the discussion of tool use by animals in Chapter 7, and of the displays of male bowerbirds in Chapter 8. But, before that, Chapter 5 looks at the evolution of animal architecture.

5

From One Nest to Another

When I was a child, dinosaurs were grey and lumbering; rather more than half a century on, they have become agile, dappled, striped, even sporting splashes of vivid colour. The speed and agility now depicted is based largely on a reappraisal of the fossil skeletons, but the colours are entirely fanciful, more the consequence of cheaper colour printing than of improved scientific understanding. So, how much fiction is there in our depiction of other aspects of dinosaurs' lives? I have a copy of a recently published biology textbook which shows a duck-billed dinosaur stooping as an attentive parent over a nest-full of 'chicks' among a small colony of other ground-nesting duck-bills. Is this vision of family life just sentimental imagining? Surprisingly, it isn't. Fossilized remnants of egg-laying sites show that duck-billed dinosaurs laid their eggs in a circular depression in the ground surrounded by a simple, low wall. Fossil embryos in the eggs confirm that these are duck-bills, and the nests are found in groups, suggesting colonial nesting.

But what about the caring parent dinosaur? Where is the evidence for that? It is based largely on the appearance of the chicks of modern birds. If I ask you to think of a day-old bird chick, what probably comes to mind is something yellow and fluffy running about pecking and cheeping—a domestic hen chick in fact. Chicks like this leave the nest on hatching and, although the mother hen is present, her role is largely one of supervision. A day-old carrion crow (*Corvus corone*) chick is utterly different. Its eyes are closed; it cannot even stand, let alone find food for itself.

There is now broad agreement based on varied evidence that modern birds are descended from a major group of dinosaurs, the Saurischia, a group that contains large predators such as the *Tyrannosaurs rex*, but also smaller oviraptors which seem to be close to the lineage that gave rise to the birds. A remarkable 80 million-year-old fossil found in the Gobi Desert reveals an oviraptor sitting on a clutch of eggs[1]—but was it incubating them or just protecting them at the moment of its sudden death? We don't know. However, the anatomy of the fossil 'chicks' of some other dinosaur species suggests that, like the chicks of crows, they could not have survived without parental care, and must therefore have been fed in the nest.

The crow's nest is a stick platform in a tree. There must therefore be an evolutionary history of nest building that leads from ground-nesting dinosaurs to the stick platform nests of the crow. But there are now over 9,000 species of birds, all apparently sharing a common ancestry, so there must also be a history of nest building that leads from a dinosaur nest to the nest of the little hermit hummingbird (*Phaethornis longuemareus*), a nest that weighs about 4g and is lashed to the end of a leaf with spider silk. The twin tasks of this chapter are to see if the history of nest building in birds, or indeed wasps or termites, can be pieced together, and whether a satisfactory mechanism can be proposed to account for the path of that history. This is a chapter about the evolution of nests, and the explanation I shall give you is what might be called neo-Darwinian: a synthesis of Darwin's original ideas, since these are the bedrock of biological thinking, strengthened by the insights of modern genetical biology. Nothing surprising then, except that, as it turns out, the way in which natural selection acts upon animal-built structures is a bit special compared with the way in which it acts upon an insect wing or fish jaw for example. If it is not obvious to you why this should be, don't worry;

I don't think it is obvious either. This is why I will explain it in some detail later in the chapter.

Swallows flying low across a meadow, jinking between the cows as they catch flies, is a pastoral image that evokes the British summer. In the United States these birds are known as barn swallows because their shallow mud nests are commonly built on narrow ledges in farm outbuildings. To scientists, they are *Hirundo rustica*. Another springtime arrival in Britain is the house martin (*Delichon urbica*), but its plump outline and less elegantly forked tail denies it such an iconic status, although its nest is more impressive than that of the swallow. It is a deep mud cup attached directly to an outside wall of your house just under the projecting eaves. The mud cup has a narrow aperture at the top where, later in the summer, the rounded faces of the chicks can be seen peering out. The sand martin (*Riparia riparia*), with its modestly forked tail and undistinguished brown colour, is the least glamorous of this summer trio but, interestingly, its nest is totally different from the other two, a burrow dug in a sandy bank overhanging a river. How did the diversity of nest design in swallows and martins evolve? This problem embodies the two concepts addressed in this chapter: evolutionary history and evolutionary mechanism.

There are eighty or so species of swallows and martins throughout the world[2] and their nest building is even more varied than that shown by the three British visitors. Driving through, let's say, the rolling hills of Brown County, Indiana, and in fact in many parts of the United States, you may see what look like painted dolls' houses on poles standing in people's yards. Closer inspection shows entrance holes, perhaps six or eight on two floors. These are nest boxes put up by bird lovers for purple martins (*Progne subis*). An alternative nest-box design consists of long-handled dipper gourds that are hollowed and then hung from a branching pole, a habit copied from Native Americans, who encouraged purple martins to nest around their villages for perhaps thousands of years. Purple martins are cavity nesters that have come to depend on human assistance, although still occasionally nesting in old woodpecker nest holes or similar

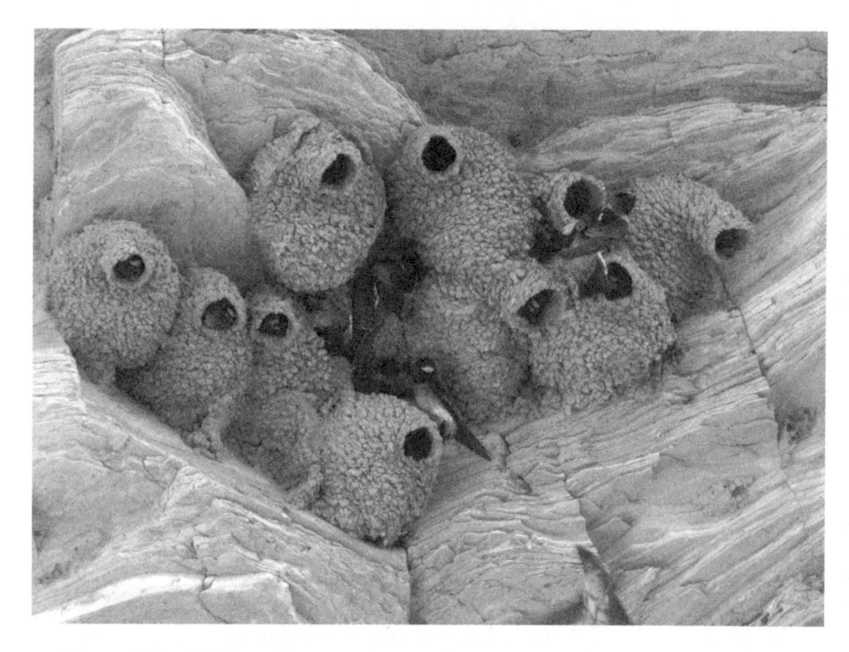

Figure 5.1. Cliff swallow nests: the globular mud nests of cliff swallows with their projecting entrance spouts represent the culmination of an evolutionary sequence in the family of swallows and martins that started with species that dug nest burrows.

Lee Rentz/Bruce Coleman Inc.

natural cavities. Cliff swallows (*Hirundo pyrrhonota*) (Figure 5.1) and red-rumped swallows (*Hirundo daurica*) build mud nests that cling to overhangs on rocky cliffs; these resemble nests of the house martin but with the addition of a narrow entrance spout. The African river martin (*Pseudochelidon eurystomina*) is a rare and local species of the Congo basin that digs shallow burrows on exposed river sand bars.

The reason that nest building in swallows and martins is diverse might of course be that they are not close relatives at all, but just look much the same. After all, swifts look and behave in a similar manner to swallows but detailed comparison proves they are a very different group of birds, closely related to hummingbirds in fact. However, careful examination of the skeletal and other body features of swallows and martins does support the view that they are a group of close relatives, and a comparison of the DNA from a selection of these species has confirmed this and allowed a probable family history to be reconstructed.[3]

Since this swallow and martin family tree is based on DNA similarities, it was constructed without any consideration of nest types. If we now sketch in the nest type against the names of the living species on this family tree, we can see the history of nest building in the group. Well, to be a bit stricter with ourselves, we see the most likely history of their nest building (Figure 5.2). In evolutionary studies of this kind, where a family tree is being constructed from comparisons between species of their DNA (or indeed any other feature, such as their skeletons) more than one family tree solution may be possible. In such a case we adopt the most economical or parsimonious explanation (i.e., the one with the minimum number of major evolutionary transitions) that agrees with all the available data. This is the principle of *Ockham's razor*, named after a remarkable medieval philosopher and theologian William of Ockham who declared that explanations should not be complicated beyond the minimum required to account for the available evidence. The 'razor' therefore refers to the process by which we pare down any explanation to its bare essentials. Later on, explanation can be elaborated as new data become available.

Back to the family tree of the swallows and martins: it seems fairly clear that the common ancestor to all the living species, the one at the base of the family tree, dug a nest burrow in the ground, as the sand martin still does. Two further nest styles then evolved from this, each represented by a number of species, cavity nesting and mud nest constructing. The purple martin and tree swallow (*Tachycineta bicolor*) are representatives of the former group, as is the South American blue-and-white swallow (*Notiochelidon cyanoleuca*), which will nest in all kinds of cavities in buildings, cliffs, trees and, as you will recall from Chapter 2, the nest cavities dug by another bird, the common miner, inside the burrows of viscachas. It is not hard to see how some individuals in an ancestral martin species might abandon the digging

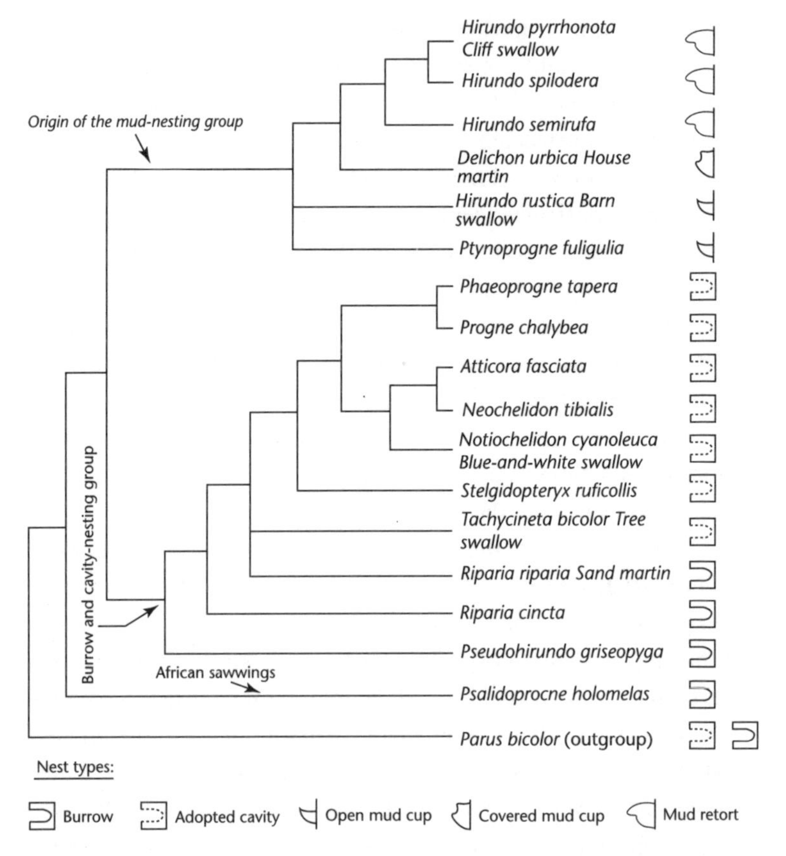

Figure 5.2. A history of nest building: a 'tree' showing the relatedness of a range of living swallow and martin species together with their nest types.

Adapted from D. W. Winkler and F. H. Sheldon (1993) Evolution of nest construction in swallows (Hirundinidae): a molecular phylogenetic perspective. *Proceedings of the National Academy of Sciences* 90, 5705–7. Figure 1

of burrows in a sandy river bank to take advantage of some other available cavities, so bringing about this evolutionary transition. After all, why go to the bother of digging a cavity if you can find a suitable, natural cavity ready to hand?

As far as the mud building group go, we can also be fairly confident from the DNA data that they all share a common ancestor, as shown on the family tree (Figure 5.2). It does not greatly stretch the imagination to visualize the diversification into the different mud built designs. An initial simple mud bracket nest, like that of the barn swallow, was elaborated firstly by a deepening of the cup to produce a design resembling that of the house martin. The subsequent addition of an entrance tube created the nest design now shown by the cliff swallow. This tube might have come about to provide special protection against weather or nest predators. What remains to be accounted for is the transition from the excavation of a burrow in the ground to the making of a mud nest on a cliff. This is not so easy to envisage.

The sand martin typically digs its burrow in a sandy bank. However, let us imagine an ancestral species digging under a bank of more clayey soil, where it also employs rearrangement of the damp earth to help create the nest cavity. Now imagine that, within the population of this hypothetical species, some individuals introduced the innovation of excavating a beak-full of damp soil of particularly good consistency from one point under the bank, then carrying it a short distance to a particularly promising ledge or crevice. Such a species could be the ancestor of the current mud builders such as the barn swallow or cliff swallow, which collect their mud loads from the edges of nearby ponds or streams and fly with it to the nest site. To make this more convincing, we need a missing link, a living species that exhibits something resembling this transitional behaviour. This species should at least excavate a cavity in a bank and rearrange the excavated material to create part of the nest wall, or even collect some mud from nearby. Such a living example we do not have. So to try to persuade you that this transition is at least possible, I am going to quickly review the evolutionary history of nest building by wasps, where the transition can clearly be seen.

In my garden every spring, tiny pale stripes begin to appear on the unpainted wood of the two patio benches. These marks are caused by wasps chewing along the grain of the wood to remove the softer fibres of the summer growth rings as wood pulp from which to make their paper nests. These social wasps, called 'yellow jackets' in the United States, are closely related to a large group of tropical and sub-tropical paper-nest-building wasps that live in colonies that are estimated to range in size from about ten to hundreds of thousands. However, careful studies on similar insects show that in this same family of wasps, the Vespidae, there are many species that as adults live solitary lives, with females building a nest and caring for the larvae on their own. Some of these solitary species dig burrows in the ground, and some build cells made out of mud. In the genus *Montezumia* there are both burrowing and building species. Mud building in vespid wasps may have evolved from either the modification of existing cavities by the addition of mud, or from the digging of a cavity or burrow to which the excavated mud is added as part of the structure. Both these routes do seem to have been taken.

The solitary wasp *Paralastor* (Vespidae) exemplifies the latter route as nest building females combine excavation and construction. They excavate a horizontal nest tunnel into a bank, but use the excavated material to build an extension to the entrance. This takes the form of a narrow tube angled upwards for about three body lengths, which then curves downwards like a hollow walking stick. At this point, the wall of the tube flares out to create a downward-directed bowl, the inside surface of which is burnished by the wasp's jaws to create a hard slippery surface.

After this specialized entrance tube is completed, the female brings paralysed insect prey to the burrow to provide food for her larvae. When sufficient prey have been brought, the female lays an egg on them and places a partition across the burrow to create a cell. She then collects more prey for a second larva and continues the sequence to provision three or four cells. After this, she plugs the burrow entrance and dismantles the walking stick tube with its upside-down bowl, leaving hardly a mark to betray the entrance of the burrow.

If you were to watch this nest entrance during the provisioning stage you would understand what task the complex structure over the burrow entrance fulfils. As the female wasp comes and goes on her prey catching trips, the burrow entrance is unguarded from parasitic insects that seek to lay eggs in the burrow from which maggots will hatch to feed on her larvae. But she has burnished the interior surface of the bowl so smooth that, as the parasites try to land in its surface, they slip and fall out again. This wasp both burrows into the soil and builds with the excavated mud. I am asking you to envisage that a similar transition could have taken place in the evolution of the nests of the swallows and martins. A cavity digging species perhaps added a mud wall to limit the access of predators. Over time, this entrance structure became more elaborate, leading ultimately to the building of the entire nest out of mud.

While on the topic of the evolution of wasp nest building, let's consider the other nest-building transition they show: the move from building with mud to building with paper. The social wasp species of the family Vespidae build nests that have small or large combs, depending on the colony size of the species concerned, and with or without an envelope enclosing the comb in a protective wall. Of these social wasps, we are confident that one sub-family of them, known as the hover wasps, represents the ancestral nest-building form. These are slender, delicate, unaggressive wasps that live in small colonies under the banks of streams or under the rocky overhangs of waterfalls, in the forests of South East Asia. Species of this sub-family show a variety of nest forms and make use of different types of material, some building nests entirely out of soil, others out of rotted plant fragments, and others a mixture of both. A plausible but unconfirmed interpretation of this is that the original hover wasp ancestor built a mud comb but, in collecting the material from the forest floor, incidentally included a proportion of rotted plant material.

Nests containing a higher proportion of plant fragments are lighter. This would have enabled them to be attached to slender plant stems rather than on rocky overhangs. Exploiting these new nest sites, new wasp species evolved that made nests entirely of plant materials, such as fibres scraped from bare wood surfaces or, as in some species, of hairs harvested from the surfaces of leaves. These materials made possible lighter, stronger and therefore larger nests, leading to the greatly increased colony sizes seen in living species of paper wasps, yellow jackets and hornets. Mud nesting species survive to this day because the mud cell walls apparently have their own advantages. Mud is heavy, but a high proportion of it in the nest wall probably prevents parasitic wasps from penetrating it with their hypodermic-like ovipositors, to lay eggs which, on hatching as little grubs, consume the resident wasp larvae. The point here is that both types of nest material, mud and paper, are adaptive in their own way, but the large colonies made possible by paper nests gave the wasps a good chance of a colony member detecting any parasitic insects before they landed on the nest. In these large colonies, the protection of the brood has been transferred from the nest material to the vigilance of the wasp workers.

At this stage I want to distinguish between two different types of transition in built structures: these are changes in *design* and changes in *technology*. In the swallow and martin example, the sequence of changes of mud nests from shallow to deep cup, and then finally the addition of an entrance spout, is a sequence where evolutionary changes are ones of design, with the material, in this case mud, staying the same. The change in wasp nest evolution from mud to paper building material is one of technology. These two kinds of change are quite distinct and, as it turns out, may have important differences in their possible pattern of evolution.

The great Palm House at the Royal Botanical Gardens at Kew in London is a cathedral of sweeping curves of glass over delicate ribs of iron. Built in 1848 to contain full-grown palms, at the time it was the largest greenhouse in the world: 363ft (110.6m) long, 100ft (30.5m) wide (that is twelve rows, each of thirteen *Routemaster* double-decker buses) and with a central vault that rises to 66ft (20m) (the height of four double-decker plus one single-decker bus), but its importance as architecture was more in the innovative use of materials than in its size as a greenhouse. The Palm House was the outcome of collaboration between Richard Turner, an Irish iron founder, and Decimus Burton, an English architect. The building is a landmark in the construction of large iron-frame structures coated in glass. This not only marked the appearance of other fine glasshouses, but also its architecture was recognized as a grand solution for the roofing of the large railway terminals being built across Britain. The single-arched roof span at Liverpool's Lime Street station of 1849 is again by Turner, this time in collaboration with William Fairburn. The mass production of cheap glass and the casting of large iron members were technological innovations, ones which then made possible new design. The Palm House illustrates evolution in architecture combining the separate elements of technology and design.

I now want to consider how a structure like a nest might evolve under natural selection, and how this process might be different for changes in nest technology and in nest design. But first, I want to set the scene for that with a diversion into the history of ideas and discovery in biological evolution.

Let us start by getting a couple of things clear. Charles Darwin was not the originator of the idea that living organisms evolved over generations, nor was he the first scientist to propose a coherent mechanism by which evolutionary change could be effected. What makes Darwin so important to biology is the mechanism that he proposed is still helping us to explain the living world. In fact, Darwin did not just propose one evolutionary mechanism but two: evolution by *natural selection* and by *sexual selection.* The first was set down in considerable detail in *On the Origin of Species*, published in 1859, that is, nearly 150 years ago, when Queen Victoria ruled an empire and the steam train was revolutionary technology. The theory of sexual selection was expounded in Darwin's *The Descent of Man, and Selection in Relation to Sex*, published in 1871. Both have proved to have immense power to explain the pattern of evolution.

The theory of natural selection proposes that the variation exhibited in the population of any species (talon size in a bird of prey, web size in a spider) is in each generation tested by the environment for its effectiveness. Individuals exhibiting the variant that is most cost-effective will leave more offspring, making that character more prevalent in the population. Suppose that selection now favours larger talon size in the bird of prey to capture a newly occurring mammal species, and a smaller web area in our spider species to specialize on the now more abundant small flies. If these selection pressures remain constant over many generations, then we will see an evolutionary trend in the prey capture of our two species as the better adapted individuals leave more offspring.

An obvious problem with this explanation is that the appearance of some animals seems too bizarre to be explained so simply and may only be shown by one sex, generally the male, with the female rather dull or at least less extravagant by comparison. Alternatively, one sex (again usually the male) may have conspicuous weapons that are lacking in the other sex. To account for these traits as 'adaptations', Darwin proposed that their context was that of sexual reproduction: weapons for males to fight off rival males for access to females, and extravagant colours, noises and movements shown by males in competition with rival males for the attention of females. This was the theory of sexual selection.

But Charles Darwin, as I have said, was not initiating the debate that substituted the idea of biological evolution for biblical creation, but contributing to one started in the eighteenth century by, among others, the Comte de Buffon, mentioned in the previous chapter (page 101). However, I need also to draw your attention to the bronze seated figure of a confident eighteenth-century man in the Jardin des Plantes in Paris. On one side of his plinth, we see him now transformed into a frail, old blind man, his hand held by an attentive daughter. This is Jean-Baptiste Lamarck, who died in poverty in 1829, his scientific achievement largely unrecognized. He is the one who can best be credited with the first exposition of a mechanism for evolutionary change. This was the theory that characters developed in parents by their activities tended to be acquired by their offspring. This notion is frequently illustrated by how it might explain a giraffe's long neck—parent giraffe ancestors by stretching for higher leaves would grow longer necks, so giving birth to offspring endowed with longer necks than they, the parents, were born with. This hypothesis tends to be scoffed at now but, coming from a man born in 1744, this was revolutionary and deserves the credit for being so. It was, in any case, another half-century after the publication of Lamarck's evolutionary mechanism in *Philosophie Zoological* in 1809 before Darwin published a coherent rival mechanism in *On the Origin of Species* (1859).

A major difference between the mechanisms of Darwin and Lamarck is that, in the former, organisms pass on to the next generation characters that they are endowed with at birth, not that they acquire during their life. This is why offspring have characters that resemble those of their parents. Darwin had no idea how the information determining these characters passed from generation to generation, but in most other key respects he has been found to be right, and Lamarck wrong.

As it happened, at the very time that Darwin published *Origin*, a monk named Gregor Mendel was beginning plant-breeding experiments in a monastery garden in what is now the Czech Republic. His experiments would reveal the rules of inheritance that were unknown to Darwin, and thereby lay the foundations of the science of genetics which ultimately led in 1953 (nearly a century after *On the Origin of Species*) to the publication by James Watson and Francis Crick of the double-helix structure of the DNA, the molecule that encodes the information that is passed from generation to generation.

What Mendel did is generally well known, but I still need to repeat some details here so that we can understand why the evolution of built structures, such as nests, has some rather special biological features. Mendel famously worked on pea plants, recording the pattern of inheritance of various plant characteristics. One of these was the texture of the mature pea in the pod, which he observed could be either smoothly rounded or wrinkled. He showed that these characters were inherited by selectively breeding lines where the fertilizing of a flower on a plant grown from a wrinkled pea, with pollen from the flowers of another plant grown from wrinkled pea, always produced a pod full of wrinkled peas (i.e. a true breeding line of wrinkled seed plants). He was also able to produce true breeding round pea plants. When a true breeding round pea parent was pollinated with wrinkled pea parent pollen, the hybrid peas that filled the pods were all of them round. However, cross-breeding between these first generation hybrids produced round and wrinkled peas in a ratio of three to one. Not an obvious result and one that has profound implications.

Using current genetical terminology, we would say that there is a *gene* or location on a chromosome that determines seed texture, but that there are two alternative forms of the gene (*alleles*—pronounced *al-eels*): one which encodes for roundness; the other for wrinklyness. An individual pea or pea plant has a pair of alleles at each gene location (*locus*), one inherited from each parent. In Mendel's experiment he started with one parent that had a pair of round-determining alleles at the seed texture gene location (by convention written *RR*); the other parent had a pair of wrinkled alleles (by convention, and we shall see why in a moment, *rr*). The first generation hybrids inheriting one allele from each parent must all in that case be *genotype* (genetic constitution) *Rr*, but their seed *phenotype* (what they look like) is all of the round type. This led Mendel to the hugely important conclusion that the round allele is *dominant* over wrinkled in determining the appearance of the seed. That is, if an individual inherits one round allele and one wrinkled allele, then only the round one influences the phenotype. A wrinkled phenotype can only occur if both alleles are wrinkled. This result is of such importance because Mendel also found that, in his pea plants, the yellow seed allele was dominant over the green seed allele, and that the purple-flowered allele was dominant over the white. Dominance was indeed a general rule of inheritance. Unfortunately for the development of evolutionary theory, Mendel, as a priest and finally as a busy abbot, did not gain recognition for his insight from leading scientists of the day. On his death in 1884, obituaries noted his work on breeding new varieties of fruit and vegetables, and his studies in meteorology.

So, all the first generation hybrids in Mendel's pea seed experiment are genotype *Rr*. This means that, as parental plants, their flowers will produce equal numbers of *R* and *r* pollen and equal numbers of *R* and *r* unfertilized seeds. As the pairing of pollen with unfertilized seed is essentially random, then the genotypes of the fertilized seeds will be *RR*, *Rr*, *rR* and *rr* in equal numbers. Because of the dominance of the *R* allele in determining seed coat, this will result in a ratio of phenotypes of three round seeds to one wrinkled. Had Darwin known this, he would have been able to say that the characteristics exhibited by a population of pea plants or hairy-nosed wombats could change over generations due to changes in allele frequencies. But what about the evolution of building behaviour? Well, Mendelian rules should apply too, provided that there are gene locations that determine building behaviour, and that the success or failure of nests can result in changes in their allele frequencies.

Pelicans have genes for fish-catching beaks, while hummingbirds have genes for nectar-feeding beaks. It is not hard to conceive how these contrasting beak phenotypes have become adapted through natural selection to their respective roles. But what about a hummingbird nest? Well, the nest doesn't actually have genes for anything. It is the product of behaviour of the bird. So does the bird indeed have genes for this behaviour, and does it make any difference to the pattern of evolution that the genes are not in the nest? This issue was raised in Richard Dawkins's (1982) book, *The Extended Phenotype*,[4] which discussed the significance of phenotypes that existed separately from the organisms that created them, the bird nest being an example of such a phenotype. *The Extended Phenotype* was a follow-up to Dawkins's landmark *The Selfish Gene* (1976),[5] which argued that units of survival over generations are not organisms but genes. In *The Extended Phenotype*, Dawkins wished to reinforce the argument of selection at the level of the gene, as advocated in the earlier volume. A bird nest helps to illustrate this because it is the phenotype, but its success or failure alters the frequencies of alleles located not within it, but within the hummingbird.

Imagine for a moment that there is a gene location that determines the kind of silk attachment for the hummingbird's nest. I say 'imagine' because geneticists get quite upset by those of an animal behaviour background, like myself, talking as if inheritance of a behaviour pattern was as simple as the inheritance of the colour of a pea flower. So my example is illustrative, rather than literal, of how building behaviour might be subjected to natural selection. Suppose this gene location has two possible alleles, one for attaching the nest to a leaf with a small amount of spider silk and the other with a large amount, and let's say that the large amount allele is dominant over the small amount one. If fewer nests fall off the leaf when attached with the former rather than the latter, then 'large amount' alleles will become more numerous in the hummingbird population at the expense of 'small amount' ones.

There is in fact plenty of evidence for genes controlling aspects of building behaviour. Laboratory bred albino mice will pull cotton wool from a dispenser to make nests; some do this more than others. Selectively breeding lines of mice for high and low cotton-wool pulling over ten or so generations (that is to say *artificial selection* as an experimental substitute for *natural selection*) leads to a gradual increase in the amount of cotton wool in the nests in mice of the 'high' selected line and to a decrease of cotton wool in the 'low' selected line. This shows that there are a number of gene locations that contribute to the behaviour, and that selective breeding over the generations leads to a gradual increase in the number of locations at which an allele promoting (high line) or demoting (low line) cotton-wool pulling is expressed.

Another example that neatly illustrates a genetic basis for building behaviour comes from the burrows of two North American mouse species. Oldfield mice (*Peromyscus polionotus*) construct a burrow system where more than one exit tunnel connects to the nest chamber. A loose plug of soil conceals all but one exit but does not prevent quick escape through any. Deermice (*Peromyscus maniculatus*), however, dig a simple chamber at the end of a single short tunnel. After twenty generations reared in laboratory cages without an opportunity to dig, mice of both species, given the chance, still dig species-typical burrows. This shows that little or no learning is involved in the behaviour; the behaviour is in essence totally inherited. The nature of the inheritance was obtained from hybrid crosses between the two species.

Hybrids of the two laboratory colonies build burrows in all respects like those of oldfield mice, while crosses between the first generation hybrids of the two species (which fortunately turn out to be fertile) produce offspring with burrow systems that exhibit varying combinations of the characters of the two species. It seems that there are a number of gene locations involved at which oldfield mice alleles are dominant over those of deermice. As a result, we are able to conclude that building behaviour can be inherited in a way that Mendel himself would have understood, and therefore that natural selection could result in their evolution over generations.

We are now in a position to go back to the swallows, martins, hummingbirds and the other nearly 10,000 species of birds, to consider how they and their nests might have all evolved from a common dinosaur-like ancestor. As it turns out, the answer may be a bit different for the nests than it is for the birds.

An accusation levelled with some justification against Darwin's *On the Origin of Species* is that on the very subject of how new species originate, it is rather unclear. Nearly 150 years on, this is no longer a problem. Modern population genetics and evolutionary theory reveal a number of different routes by which a lineage could split to form two distinct species, but the easiest one to visualize is where, at the edge of its distribution range, a species becomes fragmented into small, more or less isolated populations. Suppose the species is something like a house martin and that one of these populations begins to add a short entrance tube to the nest. This design (for reasons of local climate, nest predation or something) gives better breeding success than nests without the entrance tube. This population, due to its reproductive isolation from the other populations, develops distinctive breeding and possibly other kinds of behaviour. Later, as possibly climate change expands the distribution of the species, this isolated group merges again with the main population. However, by reason of their distinctive breeding behaviour, its members breed more frequently and more successfully among themselves than with individuals from outside the group. This leads finally to selection which reinforces the group's distinctiveness and isolates it genetically. It is now a new species, while the ancestral species continues as before; there are now two species where there was one. Both continue to evolve under the influence of natural selection, but independently.

Can building behaviour itself contribute towards speciation? Well, the evidence is that it can. The most obvious way is if there are differences in a population in the choice of nest site, which in consequence lead to the splitting of the population at breeding time. This might have indeed contributed to speciation in swallows and martins. Cliff swallows, unlike barn swallows, can attach their nests to vertical or overhanging cliffs with little or no ledge needed for support. This allows them to nest in locations unavailable to barn swallows. The family tree of the swallow and martins indicates that the barn swallow nest design was the ancestral form (Figure 5.2). If individuals within a population of barn swallow-like ancestors became able to construct nests on cliffs, they could have found themselves reproductively separated by choice of nest site from the main population. In cliff swallows, the male starts the nest with the female then joining him. In barn swallows also both male and female cooperate in nest building. Females attracted to the typical nest site of their parents, would therefore naturally pair up with the males they found at that site: cliff-nester with cliff-nester, barn-nester with barn-nester. Over generations, two distinct species could then arise.

I have hinted already that how a structure such as a nest evolves may differ in certain respects from the evolution of the nest builder itself. There is currently no clear evidence that this is the case, but it seems to me such an important matter that I want to devote the rest of this chapter to explaining the issue. I should first admit that I had never thought about it clearly until reading Richard Dawkins's *Extended Phenotype*. His book is not actually about animal building as such, but about how the expression of phenotypes can be controlled by genes that are displaced or detached from them. A bird nest is a clear example of such a phenotype but so, less obviously, is the behaviour of a host that is being manipulated by genes located in a parasite. I am going to make a little digression to talk about parasite manipulation of host behaviour on the way to talking about the web building of a spider. Let us, therefore, start with the control of a parasitic worm over the behaviour of a grasshopper.[6]

The nematomorph hairworm *Spinochordodes tellinii* (a parasite) reaches full development in the body cavity of a grasshopper or cricket. The mature worms, however, mate in water where they lay eggs from which the larvae hatch. These return to vegetation on land to be eaten by more grasshoppers. The problem for a mature worm inside a grasshopper is of course that grasshoppers neither live in water nor do they readily jump into it. The parasite needs to induce the grasshopper to do it. This it does by releasing a chemical signal into the blood system of the grasshopper, which persuades it to jump into the water. The worm then bursts out of the body of the drowning host (like the bloody emergence of the infant *Alien* from Kane's guts in Ridley Scott's (1979) film of the same name), swims away and mates. The extended phenotype of the parasite is therefore, in this case, the altered behaviour of the host. This behavioural change is mediated by a chemical, the synthesis of which is effected by genes located in the parasite.

There is in fact a parasite that manipulates the building behaviour of the host. The extended phenotype here is a built structure, built by one organism under the influence of genes located in another. The parasite in this case is a wasp larva and the host, a web-building spider.[7] The spider species in question goes by the name of *Plesiometa argyra*, which spins a fairly typical orb-web with twenty to thirty-five radii supporting a capture spiral. The web has two slightly unusual features: firstly, its orientation is horizontal; and, secondly, during web construction the spider removes all the threads at the hub to create a small open ring inside which it sits waiting for prey.

Unfortunately for the spider, not all insect visits are welcome. If the visitor is a wasp by the name of *Hymenoepimecis argyraphaga*, then the spider may end up not with a meal but with a wasp egg stuck to its abdomen. This wasp belongs to a family (Ichneumonidae) in which the larvae are parasitic and, when the egg hatches, the emerging larva begins to feed on the blood of the spider through tiny puncture holes. As the wasp larva grows, the spider remains apparently unaffected, continuing to make normal webs, feeding and growing. But, at the point when the larval parasite is fully grown, things change. The larva needs to spin itself a cocoon in which to pass the pupation phase before emerging as an adult wasp, and for this it needs a safe location. This it achieves by forcing the spider to spin a completely new structure, a *cocoon web*.

Within hours of the wasp larva starting to spin its own cocoon, the spider begins to build this special web, which is quite unlike any built by an unparasitized spider. You might expect that by now the spider is weak and disorganized in its behaviour, but instead it is purposeful and economical. It creates lines radiating from a central point but, unlike normal webs, these radii are not single-stranded but repeatedly reinforced to become sturdy cables. Also, instead of around thirty radii, there are generally only a handful of them, and they span a distance shorter than that of a normal orb web and exhibit repeated branching at their ends to create multiple attachment points. A further difference is that there is a compact arrangement of threads at the centre of this array, to create a distinctive hub. When the spider has completed this special web, the parasite sacrifices it by sucking out its remaining body fluids and discarding the husk. Then, hanging from the web hub, the parasite spins a cocoon with its own silk, and pupates.

How is this wasp larva able to manipulate the behaviour of the spider in such an apparently complex and specific way, when it is not in direct contact with the spider's nervous system but on the outside sucking its blood? The only way would be by a chemical signal injected into the spider's bloodstream. Experiments confirm this. If the parasite is removed several hours before it starts to make its cocoon, then the spider continues to live and build normal webs. The later the removal of the parasite takes place, the more the behaviour of the spider is altered. The signal is indeed a chemical one, and it is injected at quite a late stage.

The secret of how this chemical signal produces such specific and apparently elaborate alterations is revealed from the details of normal and manipulated web-building behaviour. The parasite, it transpires, does not cause the spider to produce totally new behaviour; instead the host simply misses out some parts of its normal sequence. The construction of the initial web frame and anchor lines by an unparasitized spider involves repeated adjustments in which lines are cut and repositioned, or reeled in and eaten while a new thread is paid out and attached in a different place. These behaviours are absent in the manipulated, 'cocoon web' construction. This contributes to the repeated reinforcement of radial lines to form cables and to the accumulation of multiple attachment points that are characteristic of parasitized spiders. Furthermore, the absence of hub removal behaviour leaves a central platform. The end result is a place for the parasitic larva to attach its cocoon, a place which is secure from predators such as ants, and resilient against physical damage from heavy rain or falling leaves for the duration of the wasp's pupal period.

What is the biochemistry of the drug injected by the larva of this parasitic wasp into this spider? Does it contain several active ingredients or just one? We still don't know. However, on the basis that one gene can be responsible for the synthesis of one protein, we can realistically say that it is probably a small number of gene locations in the insect larva that are responsible for the determination of the extended phenotype of the spider's cocoon web.

How might such manipulative behaviour evolve? At each gene location in the wasp larva that is involved in the production of the chemical signal there are likely to be alleles that allow variation in the biochemical synthesis. It is upon this variation that natural selection can act. The effectiveness of the biochemical components of the drug in creating a web will determine the allele frequencies at those gene locations in successive generations of wasps. The phenotype may be created by another creature, in this case the unfortunate spider, but the evolutionary consequences can be understood in terms of conventional Mendelian genetics.

Built structures, it now appears, may be the phenotypes of genes located in organisms that create them either directly, as are the nests of birds, or indirectly, as is the special cocoon web built by a spider under the control of genes in a wasp. In either case it would seem that these extended phenotypes should evolve in much the same way as the phenotypes of the organisms that cause them to be built. However, there is a problem with that conclusion. It arises when the nest, web or other structure is built by more than one individual, a fairly common occurrence.

A cliff swallow nest is, as we have seen, not the work of one bird but the joint effort of a male and a female. So the extended phenotype represented by the nest is the product of genes present in not one organism but a pair, and that pair are genetically different. The mound of a termite is the product of the collective efforts of perhaps 100,000 individuals that are not genetically identical. Does this make the nature of evolution of cooperatively built structures different in any way from the evolution of the builders themselves? A mature mound of *Macrotermes* termites may contain 3 or 4 million individuals but they are all the progeny of a founding pair, the queen and king. Since the method of inheritance in termites is just like ours, these brother and sister workers will show the same range of differences in behaviour (that is, differences in their phenotype) as do human siblings. You'll probably know from personal experience that brothers and sisters do not always agree. So how do these sibling termites agree on what to build? This is where the distinction I made earlier, between changes in technology and changes in design in the evolution of structures, becomes particularly important because I believe the solutions may differ for these two cases.

Let's start with the technology problem and let me précis Richard Dawkins's 1982 (*The Extended Phenotype*) explanation for how natural selection might bring about a change of building material in mound building by termites. Imagine a colony of hypothetical termites in which choosing dark or light mud as a building material is determined by a pair of alleles at one gene location, and that the allele for selecting dark (which we will label *D*) is dominant over the allele for selecting light material (labelled *d*). If both the queen and king are of genotype *Dd*, they will both produce equal numbers *D* and *d* gametes (eggs or sperm). Pairing randomly, these will produce equal numbers of offspring with *DD*, *Dd*, *dD* and *dd* genotypes, producing a ratio of behavioural phenotypes in the workforce of three to one in favour of choosing dark mud. As the nest is built up from many thousands of mud loads contributed by a huge workforce, the mound will be a blend of dark and light in these proportions. For the composition of the mound material to change over generations through natural selection, we can imagine that a predominance of light material in the mixture is better (less easily eroded by rain, perhaps). The result is that mounds with more pale material (that is, ones with a higher proportion of *dd* individuals) produce more new queens and kings which, like the workers, are simply the offspring of the original queen and king. This new generation of nest founders are more likely to be carrying *d* alleles than the dark nests which are producing fewer offspring. The result is an increase over generations of the proportion of *d* alleles in the termite population relative to *D* alleles.

In case that seems a bit abstract, consider again the nest material of South East Asian hover wasps. As I mentioned earlier in the chapter, some species of these wasps build nests wholly of mud, some only with decayed plant material. In some species, however, both materials are present, although the proportions may vary between nests, the blend in any nest depending on the number of colony members collecting each sort of material. As we have already discussed, both mineral and organic materials have their merits and their limitations. Mud is heavy but may prevent parasitic wasps laying eggs on wasp larvae through the nest wall. Organic building materials make light nests that can be attached to a wider range of locations.

So what proportion of mud to organic material is best? Well, that depends on the local selection pressures. Where the level of parasitism of the larval brood is high, selection should favour nests where a high proportion of colony members collect mud; where parasitism is absent, nests where colony members collect predominantly plant material should benefit. Long-term changes in local selection pressures should lead to an evolution in the predominant building material collected by any colony, i.e. to the evolution of nest technology. Notice that, even if colony members have different individual preferences in building materials, the nest ends up being a blend of their collective efforts, and is adapted to local selection pressures. There does not seem to be any obvious reason for conflict or confusion in a colony where different colony members are choosing different materials. Therefore there is no problem in understanding how what I am calling a technological change (a change on nest material) might occur through evolution, even in collectively built structures.

But what about evolution in the design of collectively built nests? If different colony members are trying to build a different structure, will there not be conflict or confusion? Richard Dawkins realized that this is harder to predict than disputes over technology, but suggests there might be, for his hypothetical termite colony, resolution through a voting system in which the majority prevails. This of course requires the termites to convey their wishes to one another. Although it would be cumbersome, this is a reasonable notion. As we have seen in Chapter 4, social insects communicate and, in honeybees for example, there is a consensus-based communication system for deciding whether a potential new nest cavity is suitable for the colony to move into. A similar consensus system is used by a colony of the ant *Temnothorax albipennis* to decide upon the most suitable of a choice of potential nest cavities. However, consider a dispute over the building of the royal cell for a queen termite.

In the classic experiment of Pierre-Paul Grassé, described in Chapter 4, on the rebuilding of a royal chamber for a queen termite, the position of the wall of the chamber, you will remember, was determined by the threshold value in the gradient of the cloud of *pheromone* or chemical signal that she produced around her body. But there is a problem with this explanation when one considers that the wall results from the collective response of the workforce. What happens where the workers differ in the genes that they carry for the threshold value? In this case, if workers with a lower than average threshold value build a wall a certain distance from the queen, workers with a higher than average threshold value might add material to its inside surface, so producing a thicker than necessary wall. It seems possible that high threshold individuals would succeed in doing this, even if in the minority. All would cast their vote, but the will of the majority would be thwarted.

Currently we know nothing about either the extent of design disputes within a termite colony, or how they are resolved. However, the *lattice swarm* models of nest building by virtual wasps that we were also discussing in Chapter 4 (Figure 4.1) seem to provide an example where the impact on nest architecture of dissenters in the workforce would be minimal. In lattice swarm models, you will remember, virtual colony members move around a three-dimensional space, adding standard building bricks when they encounter specific local architectural configurations. The purpose of the models was to demonstrate, as they did very effectively, that complex structures could be produced by a workforce that did not communicate with one another and responded only to local stimuli using simple behavioural responses.

In the lattice swarm models all colony members were assumed to share the same rules. The question that we are asking now is if colony members were not all equipped with the same building rules, how could architectural catastrophe be avoided? The lattice swarm models suggest that the nest structure itself might not permit it.

The lattice swarm models revealed the existence of two types of algorithms or sets of building rules, coordinated and uncoordinated, the great majority being of the latter type. Only coordinated algorithms repeatedly and reliably generated the same nest architecture each time they were run, and the architecture they generated was characterized by a certain coherence or orderliness; indeed it showed a modularity resembling that seen in the nests of many wasp species. However, coordinated algorithms had another property that is particularly important where members of the workforce differ in their building rules: the architecture they generate is very little affected by the addition of random behavioural rules. The reason for this is that a rogue builder equipped with such an additional rule can rarely find a configuration in the virtual nest to which the random rule is applicable. Coordinated lattice swarm algorithms turn out to be very robust, the constraints of a nest's architecture limiting the ways in which it can grow.

This property of robustness in the architecture of a growing nest allows some speculation on how wasp nest architecture might evolve. Similar yet distinctive architecture is known to be generated by related coordinated algorithms. This suggests that a few specific building behaviour mutations accumulated over generations might be able to bring about the transformation of one nest architecture into another. The problem is the potential disruption to nest architecture during those transitional generations, where builders may have inherited some but not all of the necessary behavioural changes. But the robustness of coordinated algorithms seems to indicate that this disruption might be minimal. Perhaps a new nest architecture could appear suddenly in a single generation when two appropriate parents produced offspring with the genotype required for the new coordinated algorithm.

The evolution of structures built collectively does raise some special questions about the nature of the evolutionary process. Scientists are explorers. They are attracted to unknown territories. This seems like one worth an expedition.

6

Two Routes Lead to Trap Building

I have a fish trap that I bought in a market in Thailand a few years ago. It is sausage-shaped, 60cm long, 12 wide, and made of fine strips of rattan palm. It is really a highly specialized basket in the form of an open lattice cage, permanently closed at one end and stoppered at the other. Into the side, at the end with the stopper, an ingenious valve has been fitted; a narrowing funnel of springy rattan teeth, through which a fish can force its way in but which bar its way out. It is a variant of traditional fish traps made of natural materials which can be found the world over: a baited cage with a valve entrance. A stoppered (sometimes hinged) door is there for insertion of bait, and later for the removal of deluded victims. The elements of these traps are that they attract, trap and then restrain the prey while unattended by humans.

I bought my fish trap because of the workmanship. It is a beautiful example of basket-making. Arranged lengthways, there are about 115 strips of material held in position by twenty-six hoops of finer bindings, creating an open lattice with a mesh of about 212 × 2mm. I estimate that the number of fastenings needed to create the trap is over 2,500. That is quite a bit of skilled work.

It is difficult to be sure when in human history such traps first appeared. They will have perished quickly, leaving only indirect evidence. There are detailed illustrations from ancient Egyptian tombs of nets being used to trap fish and also wildfowl. However, that only takes us back three or four thousand years and, from the textiles preserved around mummified bodies, it should not surprise us that the Egyptians could also make nets. I have seen the invention of textile manufacture referred to as the 'string revolution'. Certainly it must have been enormously important for humans in making all kinds of objects, including clothing and also net traps. But when did it occur? You are probably familiar with the 'Venus' figurines found in a few Palaeolithic sites in Europe, their bulging bodies appearing more obese than pregnant. These date from around 25,000 years ago. An odd feature of them is that they lack faces, the whole head sometimes being covered with a regular pattern of criss-cross lines. These lines have been interpreted as elaborate hairstyles, but a recent alternative suggestion is that they are hairnets or snoods. A few of the figures indeed seem to be wearing textile skirts of some kind. Perhaps these people also made fishing nets.

Humans do make and have traditionally made all kinds of traps to capture animals, in the water, on land and in the air. All have a degree of complexity: multiple fastenings, moving parts, valve system or use of bait. How simple can a trap be? One of the simplest must be a snare. This is another almost universal trapping device, used in Africa, for example, for the capture of 'bush meat'. It consists of a piece of wire with an 'eye' at one end through which the other end is threaded and tied to a low branch or sturdy stake. There is no bait; capture depends upon the prey animal putting its head into the wire loop and pulling it tight. The more the victim struggles, the tighter the noose gets. The probability that the intended victim will oblige depends on the skilful choice of location and arrangement of the snare. When mammal head and wire noose have met up, the success of the trap depends upon the smooth flow of the wire through its 'eye', the strength of the eye itself and the strength of the fastening at the other end to prevent the prey escaping.

So the snare is a device that can be reduced to one material, two fastenings and no bait, compared with my fish trap with its 2,500 fastenings plus use of bait. Even so, can you imagine next year a group of chimpanzees being found in the Congo that set snares to catch prey? What would you say if asked to write a short item about this for a newspaper? What headline would you demand, and how big? Ourselves apart, no living primate makes a trap, nor indeed any non-human mammal at all. There are no trap-making birds either, although there is now some evidence that at least one bird species uses bait to attract prey. The bird is the burrowing owl (*Athene cunicularia*) and the bait is mammal dung. This is worth explaining since it illustrates the extent and limit of the trapping powers of birds.

Burrowing owls, as was mentioned in Chapter 2, nest in burrows made by rodents. They don't make any nest as such, but a puzzling aspect of their behaviour is a tendency to gather cow dung and place it in or around the burrow entrance. An experiment was conducted to test between the two rival hypotheses:[1] (a) that the smell of the dung is used to conceal the smell of the nest from predators; (b) that the smell of the dung is to attract insects on which the owl can feed. Notice by the way, that these are not mutually exclusive; it is possible for results to support both. In fact the nest-concealing hypothesis is not supported. Fifty artificial nest burrows were dug by the experimenter and baited with domesticated quail eggs. Twenty-five of these burrows received the cow dung treatment and twenty-five 'control' burrows were left without. The result? Experimental and control burrows were equally predated, and equally rapidly.

The insect bait hypothesis on the other hand was supported. The regurgitated pellets of food remains that were spat out by owls supplied with additional cow dung had ten times more dung beetle remains than those of 'control' owls that received no cow dung supplement. So here is evidence of the use of bait but not of trap construction, just of direct predation. Not only is there no trap building in non-human mammals, but none in birds, nor indeed in lizards, frogs or fish. There were five hundred million years of vertebrate evolution before the first trap builders came along—us.

What about invertebrate trap builders? Well, that's different. There are certainly thousands of species of them. You only have to think of the spiders, for which trap building is almost a defining feature, to realize that there is a mass of possible examples. Of course, we talk of spider 'webs' rather than 'traps', but traps are what they are. All the principles that I have been admiring in my fish trap: precision manufacture, moving parts, immobilization of prey, bait, site selection and trap orientation can be found in spider's webs, although not necessarily all in one type.

The distribution of trap making across the animal kingdom therefore looks very odd: numerous invertebrates and us. Yet our response to this seems to be remarkably lacking in curiosity. The discovery of any vertebrate trap builder other than ourselves would be a great surprise to us. Nevertheless we accept as unsurprising, albeit ingenious, that spiders place their webs on and around our houses to trap prey. The spider and ourselves are both employing the same principles in our traps and both achieve the same goal. Why our contradictory attitude? They achieve the goal in a quite different way to ourselves, yet we fail to articulate what that way is. Let us remind ourselves of the primary constraint on invertebrate building behaviour: small brains. This chapter is an assessment of whether there are two distinct evolutionary routes to trap building: the big-brained way, that is our way, and the small-brained way for all those invertebrate trap builders. Also, if there truly is a small-brained route, what is its evolutionary path?

As a scientist it is important to check the premise upon which your argument rests. Maybe spiders are actually the invertebrate exception? They do have complex and sophisticated behaviour, the organization and control of which has been somehow miniaturized. Let us first return to the theme of Chapter 3, the simplicity of construction behaviour. Does spider's web building fulfil the predictions made there of simple, stereotypical, repetitive behavioural elements?

Figure 6.1. ***Araneus*** **orb web: this spider's orb web consists of a capture spiral placed on a radial array of threads; this is supported within a frame thread that is attached at several points to local plants.**

David Ponton/Getty Images

There was also the supplementary prediction that such complexity and flexibility as is evident in building behaviour should be seen in the behaviour of getting the structure started rather than in the main construction sequence. All these features are nicely illustrated by the building of the typical spider's orb web (Figure 6.1).

One of the most studied species of orb web spider is the 'garden' or 'garden cross' spider (*Araneus diadematus*). Ignore for the moment how it creates the web frame enclosing the array of twenty to thirty radial threads that converge on the central hub. How does the spider place the spiral of thread that bears the sticky droplets to catch the prey on to the radii?

The spider, starting from the hub of the web, first spirals out towards the frame of silk threads inside which the radial threads are arranged. At this stage it is not laying down the capture thread but a 'scaffolding' thread. This it joins to each radius in turn, using the rule that the threads intersect at the same angle on each occasion. It is not clear how the spider manages to do this, but the effect is to create a spiral in which the distance apart of each turn increases towards the edge.

The 'scaffolding' thread is so called because it is a temporary structure put in place to stabilize the radii during the laying down of the capture thread spiral, the thread that will actually trap the prey. This is laid down by the spider starting at the edge and spiralling back towards the hub, at the same time picking up the scaffolding thread. The rule the spider uses here is that each turn of the thread is equally spaced, and we know that it does this by the simple method of holding on to the previous turn of the capture spiral with its 'outside' front leg. This provides a simple measuring stick for the point of intersection between the capture thread and each radius. The elegant arrangement of the capture spiral, for all that it creates a beautifully geometrical effect, is accomplished by simple, stereotyped, repetitive actions.

How the spider constructs the frame and radial threads in the first place is more interesting, particularly bearing in mind that, before starting the web, the spider will frequently have incomplete knowledge of possible web attachment points. However, from watching the behaviour of an orb web spider, it is evident that it makes use of a consistent set of tactics, although these emerge from a variable sequence of actions.[2]

In its natural habitat a spider uses gravity to create a vertical thread, attaching it at one point, and dropping down while secreting a silk line, then attaching it at the bottom. On reaching a firm footing, the spider may, however, move to one side, paying out more thread before tightening and attaching it; this way it can create an oblique thread. A spider placed on an isolated twig may try to make contact with another attachment point by 'floating' a thread on the breeze until caught on another solid object. The spider then pulls the thread tight, travels across on the thread, laying down a more secure replacement. From this bridge, the spider may then explore downwards with the aid of gravity.

A spider is not dependent upon thread floating to create a horizontal thread. If an *Araneus* spider is placed on a simple U-shaped wooden frame, it can put in a horizontal thread to enclose the top side of the frame by attaching the thread at the top of one arm of the 'U', walking down, along the bottom and up the other side, paying out a silk thread, which it then tightens and attaches at the top of the other arm. Repeatedly attaching a thread, moving to another point while paying out silk, tightening and attaching the thread again, allows the spider to arrange threads at any angle. Also, the structure can be continually edited by the removal of threads. In this way a web frame enclosing radial threads can be built up. The ability of the spider to make use of a 'detour' walk to get between two points is key to the success of its getting started sequence.

I personally would like to see more research on the flexibility of spider behaviour in the opening stages of orb web building, given, for example, the various combinations and arrangements of stems and branches with which a spider may be faced. There could be some interesting findings, although I don't expect to see any evidence of a spider brain with capacities outside those we already know for invertebrates. More likely, we will find an ingenious solution to a problem that we regard as complicated.

So spiders do show some behavioural complexity in the construction of their traps. However, there is something beyond the skill of making a trap that we give ourselves as humans credit for, that is the intelligence to imagine a trap. That requires the understanding that to capture prey you don't need to lay hands on it directly, but can create a structure which acts as your proxy. Do spiders imagine the trap before they build it?

A tiny *Araneus* spiderling, isolated at birth, will build a typical orb web. To create the essential structure, it does not need to see other webs, or practise making its own. Another spider, *Zygiella x-notata*, builds a slightly different orb web design. At first glance it looks exactly like that of *Araneus diadematus*, until you notice that one of its radii goes from hub to frame without being touched by any part of the capture spiral. In other words there is a gap of two segments of the capture spiral that leaves one radius free. In *Zygiella*, the spider rests, not at the hub but at the edge of the web, in a position where it can touch the free radius. Any disturbance on any part of the web sends a vibration from the hub via the free radius to the hidden spider, which rushes out to kill the anticipated prey. This web design is an inherited characteristic of the species. Its 'invention' is apparently some ancestral mutation or mutations that caused part of the capture spiral to be omitted. The spider's orb web and its variants are the creation of the evolutionary process, passed on in the genes, generally speaking one web type for any one species of spider. There are now lots of different web types but spread among a number of species. That is very different from ourselves where a fish trap was the product of a creative mind, passed on down the generations by learning, and where, having grasped the concept of a trap, we can translate that into a host of different trap types, each adapted to local prey and building materials. This comparison supports the 'two routes' viewpoint: i.e. that the spider produces a trap through fundamentally different thought processes from those of humans.

A variant of the snare trap, and one beloved of Hollywood, has a noose or net lying on the ground but attached to a bent-over springy tree, so that once the trap is sprung the net is released, jerking the victim into the air. There are many variants of this principle. Remember, for example, in *The Return of the Jedi*, Luke, Han, Chewbacca, plus robot companions being scooped up in a net trap set by Ewoks!

Humans set traps for all kinds of prey: 'bush meat', fish, even termites. In the folk tales of the Azande people of Southern Sudan, compiled by the great twentieth-century anthropologist E. E. Evans Pritchard, there is a rascally hero called Ture.[3] His name translates as 'spider', apparently for his clever behaviour, but his chief activity is tricking friends and relatives out of hard-earned food to satisfy his own gluttony. One of his favourite meals is a porridge made from ground-up termites; these are collected by night trapping using lights.

New termite colonies are produced by the release from a mound of large numbers of winged males and females. After a single short flight, they pair and, on landing, each freshly united queen and king will dig a small cavity in the ground in which to produce their first generation of workers. In the country of the Azande, the typical pattern of winged termite emergence is at night after heavy rain. It is then that the wall of the mound is soft enough for workers to open up exit holes, and the ground soft for new royal pairs to dig themselves in before daybreak. A cloud of winged termites will emerge under these conditions, flying up towards the moon, being carried a little way on the breeze before landing, shedding their wings and digging in.

The traditional Azande trapping method is, during the day, to clear ground around a large mound where they then dig some small pits. At night after rain, when the winged termites stream out on to the surface of the mound and take flight, torches are lit. The termites, attracted to the flames, get singed, fall to the ground and, dashing about in their confusion, fall into the pits where they can be scooped up into a bag. Plump with flight muscle and reproductive organs, these make a protein-rich food either eaten fresh or dried, and ground to make termite flour. To catch termites, the Azande use a light-baited, pit trap.

Many of the trap types created by humans have equivalents in the animal kingdom. It is yet another example of there being a limited number of good solutions to a problem. The pit trap principle is rare in non-humans, being confined to two types of insects, so-called 'ant lions' and 'worm lions'. The ant lion pit trap resembles nothing so much as a tiny, cone-shaped crater. These can be seen in hot climates, where collections of them pock-mark patches of dry, dusty ground, under a tree or rocky overhang. The ant lion lies concealed under the floor of the crater. This can be demonstrated by using the tip of a grass stem to dislodge a small cascade of dust from the lip of the crater. This flows surprisingly easily to the crater floor, whereupon particles erupt upwards like a volcano, while the walls of the crater keep sliding back towards the bottom. Had that been an ant rather than your grass stem, it would have gone slithering down into the crater where a pair of huge jaws would have flashed into view to drag the ant to oblivion.

An ant lion is actually the larval form of a winged adult insect belonging to the order Neuroptera (nerve winged). This name reflects the fine network of veins typical of the adult insects of this order, which also includes the widely occurring lacewings. A worm lion, the other insect pit trap builder, is a fly larva; a very different creature, but one that makes craters that are virtually indistinguishable from those of ant lions, and which operate in exactly the same way. These pit traps, of both ant lions and worm lions, are not typical of traps in that they do not work unless the owner participates by throwing sand from the bottom of the crater, and they are not typical of invertebrate trapping methods which almost universally make use of self-secreted materials—materials that are of essentially two types: silk and mucus. Of these, silk is the more important, but let's start with mucus.

There are a number of slimy, sticky secretions produced by invertebrates that are referred to as mucus although their biochemistry is not well studied. Much better known are mucus secretions in humans. These can be found lubricating the lining of your gut to prevent it being scratched by passing food, and as lubricants of your limb joints. The essential constituent of these mucus secretions is polysaccharide. Molecules of this kind are frequently found arranged as an array of long, unbranched chains attached to the sides of a core protein molecule. These complex molecules, called *proteoglycans*, have a tendency to bind water, which assists their role as lubricants.

For slugs, mucus is the material that allows them to lubricate their movement over your lettuce leaves, for earthworms it lubricates their progress through the otherwise abrasive earth of your lawn. But for other species, the high viscosity of mucus is a property that, combined with its stickiness, offers the potential for making traps especially in water. We have already seen this in the mucus house of the *Oikopleura* in Chapter 3, complete with its very fine filter nets of mesh $0.3 \times 0.3\mu m$ (micrometres) for the capture of minute food particles (Figure 3.4). You will remember that it belongs to the Phylum Chordata, which is totally dominated by the vertebrates, but the important point in the trap-building context is that *Oikopleura* has adopted the classic invertebrate route to trap building: simple construction behaviour and self-secreted materials.

Looking at not just the fineness but also the regularity of this mucus net in *Oikopleura*, it seems extraordinary that the material can achieve this by just being stretched. The closest I can come to an analogy, and I feel slightly embarrassed at making a connection with something so clumsy and banal, is to remind you of what happens to the mozzarella cheese when you try to separate your pizza slice from the mother-disc. The point is that all you do is pull; the array of cheese strands just get finer as they continue to span the growing distance between the two pizza pieces; it is a property of the cheese. *Oikopleura* not only produces nets of mucus to cover the water inlets and to filter out food particles, but also mucus for the enveloping wall of the house, showing that it can produce different sorts of mucus for different specialized jobs—clever, self-secreted materials.

The burrow-dwelling marine worm, *Praxillura maculata*, also uses a mucus net to capture fine particles suspended in the water. As support for the net, it first builds above its burrow a hollow tube at the end of which, in a manner not yet studied, it adds a star of fine radiating spokes, curving slightly away from the tube mouth. On to this scaffolding the worm secretes a fine film of mucus like the fabric of a delicate, diaphanous umbrella (Figure 6.2).

Some species of mollusc have mimicked the evolutionary path of the segmented worms in adapting mucus secretions that formerly assisted locomotion to trap food. The result, in one case is a mid-water, oceanic, plankton-feeding, shell-less snail by the name of *Gleba cordata*. It drifts in the current, deploying a large, fine-meshed mucus net one third of a metre or more across. As the net drifts, microscopic animals and plants collide with it and stick to its surface.

We don't have detailed evidence of the complexity of the behaviour involved in making these various mucus nets, but there is nothing

Figure 6.2. Marine worm web: the mucus capture net of the marine worm *Praxillura* is held by six spokes that radiate from the mouth of the dwelling tube.

Photograph by McDaniel and Banse

to suggest that it is anything other than simple. It conforms exactly to the small-brained building solutions seen in Chapter 3: complex architecture is possible with simple minds and clever materials.

As a material for building nets, silk is in certain respects greatly superior to mucus. It is elastic, that is, it recovers its former shape after deformation, rather than viscous, making silk nets more durable than mucus ones. It is also much stronger, allowing the capture of much larger prey. It can be spun in water, as it is by the larvae of caddis flies to make capture nets, and also on land, as it is by spiders to make webs. Spider silk is certainly a very clever material, but I'm going to start with the simpler net traps of caddis.

Caddis fly larvae are probably better known for building portable cases, as described in Chapter 3, than traps, but a minority build shelters attached to the bed of a stream where they also construct a net to filter food from the current. Different net mesh sizes are used by different species, larger mesh sizes in faster water currents catching larger food items, and smaller mesh nets capturing smaller food particles suspended in slower moving water. The caddis larva *Macronema transversum* is a slow-water specialist that builds an integrated house and filter net. Sand grains are stuck together with silk to form a rounded capsule in which the larva lives. Water is encouraged to flow through this house by the inclusion of a sand grain funnel that faces into the current. Filtering of the water flowing through the house is achieved by the placing of a net curtain of silk with a fine mesh across the internal cavity from which the larva can grab trapped particles. *Parapsyche cardis* is a fast water specialist; the mesh size of its net is about a hundred times larger than that of the smallest mesh caddis nets and about a thousand times the mesh size of the mucus feeding net of *Oikopleura*. When you consider also that the mesh size of the web of a modest-sized spider is a few millimetres side length, the full-size range of capture nets made by invertebrates from self-secreted materials becomes apparent. The largest has a mesh size about 10,000 times bigger than the smallest, and consequently such nets are used to capture a very wide range of differently sized food items.

The nearest relatives of the caddis larvae are the caterpillars of the butterflies and moths. Essentially all of these secrete silk, yet none has evolved a capture net nor indeed any other kind of trap. The most obvious difference in their way of life compared with that of caddis larvae is that the latter are almost wholly aquatic, while caterpillars are almost universally terrestrial. Air is so much less dense than water that small, potential food particles do not remain suspended in it long enough to provide a living for filter net feeders. Filter nets, whether the silk ones of caddis larvae or the mucus ones of *Oikopleura*, are only suited to water.

What air does contain is flying insects which, because air has such a low density, can readily determine their own direction of movement rather than simply being carried on the currents. To capture these generally requires not a filter net, but a net that will intercept flying prey and then prevent them from flying away; the mesh sizes of these are very much bigger than those of the aquatic filter nets. Spiders, and almost exclusively spiders, have managed to evolve such capture nets. The caterpillars of butterflies and moths, for all that they are adapted to produce silk, sometimes in large amounts, have only evolved the ability to build larval or pupal shelters with it.

Why can't caterpillars build webs like spiders? That is a matter of speculation, but remember that spiders, even before any had evolved capture webs, were predators. The modern tarantula spiders, which build silk-lined tunnels or retreats, show this trait. Caterpillars are almost exclusively herbivores. They would have needed two major evolutionary changes to compete with web building spiders in the capture of insects by using webs: the first, to evolve carnivory, the second, to evolve webs. With the spiders one step ahead of them, it seems caterpillars have never been able to compete. A few exceptional caterpillars have in fact become carnivores—about one species in a thousand in fact. The genus *Eupithecia* (the 'pugs') has several hundred species worldwide of typical, leafeating, 'inchworm' caterpillars, but a handful of them on Hawaii have become ambush predators, grabbing and eating unwary insects that venture within their reach.

The case-bearing caterpillar of another Hawaiian moth species *Hyposmocoma molluscivora*, as its species name indicates, is an eater of snails. It also uses silk during its attack, not to trap the snail but prevent its escape. This is not some speedy snail that can outrun the caterpillar, but it could escape just by dropping off the plant on which it is typically found, were it not that the caterpillar, on discovering it, quickly ties it down with silk. The caterpillar then forces its way under the rim of the shell and, leaving its case temporarily attached to the outside of the snail shell, eats its way further and further inside.

Given the failure of caterpillars to emulate the spiders in producing silk traps to capture airborne prey, it is surprising that a species from a quite different order of insects not noted for their silk production, has done exactly that. The insect is a fly or, more precisely, the larval stage or maggot of a fly called *Arachnocampa luminosa*. It is found living on the ceilings of the dark interiors of caves in New Zealand. Its trap consists of twenty to thirty hanging silk threads, each coated with a mucus-like secretion which forms into a string of sticky beads.[4] The fly larva lies in wait in a silken retreat from which the capture lines hang. But the larva is also using a bait, and the bait is light. At the posterior end of the transparent body of a larva is a light-producing organ. Looking up in the darkness at the roof of the cave is like looking at the night sky studded with stars. These lights attract small flies that, flying upwards, strike the hanging threads where they become stuck, are reeled in and eaten.

Silk threads and sticky surfaces are exactly the special materials that are the basic components of spider traps. Spiders have in fact used these to evolve a wide variety of prey capture devices, not simply the well-known orb type, each designed for the capture of a particular type of prey. The Australian, redback spider (*Latrodectus hasselti*), a relative of the black widow, builds a trap designed to capture walking prey; the spider is a specialist predator of ants. The bad reputation of this spider comes from its powerful, occasionally fatal, venom, and from its liking for living around people's houses. Its web is a loose tangle of threads which might be located under some piece of garden furniture. From this, silk lines are dropped to the ground and secured under tension, a few sticky droplets are added to each of these lines just above the ground surface. When an ant dashing along the ground hits one of these sticky droplets, it struggles, breaks the thread attachment and is jerked into the air where, dangling helplessly, it can be dispatched at the spider's leisure—the capture principle embodied in the noose attached to a springy tree.

A similar capture principle is used by a spider that specializes in the capture of 'pond skaters' (family Gerridae), insects that are light enough to stand on the surface film of the water and scud across it with sweeps of their legs. Pond skaters are predatory, seeking out the ripples made by struggling prey trapped in the water film. However, an unwary pond skater may find itself trapped in the web of a species of *Wendilgarda*, a loose bead curtain of silk threads coated in sticky droplets that hangs down into the water from overhanging plants[5]—the insect predator now turned spider's victim.

More than any other, the design that has come to symbolize spider trap construction is the orb web. This, or rather I should say *these*, since there are various, species-specific variants of the orb web, are designed to capture flying prey. We have already checked the possibility that building a structure of such elegant design required particularly complex or sophisticated behaviour, and it didn't. It is now time to check whether the real sophistication of the orb web builder lies in its self-secreted materials, and it does.

Silk is a term applied in a rather general sense to threads extruded from the glands of a variety of invertebrates. These secretions are more often than not used to make shelters, as in the making of pupal cocoons by the caterpillar of the silk moth (silkworm), *Bombyx mori*, from which most commercial silk yarn is obtained. The biochemistry of silk varies across the arthropods, but silkworm silk is a protein made up of long chains of *amino acids*. A few amino acid molecules linked together are referred to as a *peptide*, and a longer chain of amino acids as a *polypeptide*. A polypeptide is essentially part of a protein molecule.

The silk of spiders is very similar in composition to that of the silkworm, so I am going to spend a bit of time explaining the biochemistry of silk. If you never liked chemistry or have never done any chemistry, don't worry. This is not really a chemistry lesson at all; it is still about architecture and engineering, but using bricks the size of molecules. The wonder of the orb web is that we can see and understand how it is engineered to capture fast moving prey—engineered at the level of the whole web, the level of the individual thread, and at the level of the protein molecule.

When I say that silk protein is made up of chains of amino acids, don't think of these chains as necessarily straight. They can be bent into three-dimensional shapes, sometimes in highly organized ways. At the heart of an amino acid molecule is a carbon atom. It can bond with four partners, one to each of its four 'arms'. Three of these partners are always the same; one is variable. The three invariant partners are: a *single hydrogen* atom (H); a *carboxyl* group (COOH: carbon, oxygen and hydrogen); and an *amino* group (NH_2: nitrogen and hydrogen). For the variable partner there are about twenty options (which is why you need varied sources of protein in your diet, to ensure the intake of a full range of amino acids). Some of these variable groups are quite large (containing strings of up to four carbon atoms and associated hydrogens, or rings of five or six carbons e.g. tryptophan), but the simplest of these variable groups is a single hydrogen atom. The name of the amino acid with this basic conformation is called *glycine* (Figure 6.3). The next most simple amino acid has a variable group of one carbon and its three associated hydrogens (CH_3). That amino acid is *alanine*. So the key message of this chemistry lesson is that glycine and alanine are the smallest, most compact bricks in the whole family of amino acid building blocks. Not by coincidence, the most important amino acids in the spider's orb web are glycine and alanine.

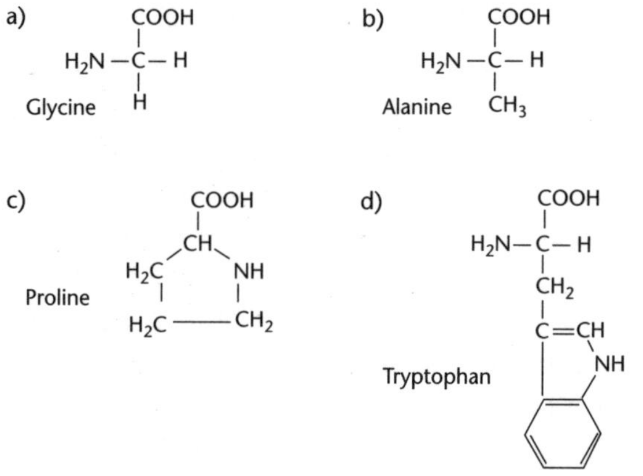

a) Glycine: $H_2N-CH(COOH)-H$

b) Alanine: $H_2N-CH(COOH)-CH_3$

c) Proline

d) Tryptophan

Figure 6.3. Amino acids in spider silk: the most simple of all amino acids, glycine and alamine, are also the most important in spider web silk.

An orb web spider such as the garden spider *Araneus diadematus* extrudes silk as a liquid from spinnerets at the posterior end of its abdomen. The silk instantly hardens as a fine thread, the thickness of which is defined by the size of the aperture through which it came, i.e. defined by the anatomy of the spider, not by its behaviour. Similarly, the composition of the silk is a property of the gland, not of the spider's behaviour. *Araneus* has in fact a bunch of spinnerets at the tip of its abdomen associated with not one but seven different glands, each producing a distinctive, specialized secretion, five of which are concerned wholly or largely with web construction. The web therefore is made up of different specialized components, each of which makes use of specific, tailored secretions.[6] The orb web is a composite piece of engineering.

It is worthwhile pausing for a moment to consider what the orb web of *Araneus* is trying to trap, and how. Its prey are fast-moving insects of a weight that could be approaching or even greater than that of the spider itself. Such a prey item needs to be brought to a halt and then held fast by the web long enough for the spider to dash to it and deliver a paralysing bite. Failure by the web to absorb all that energy of movement, and the insect will pass right through the web and escape. The web's capture surface may be about 15cm across, but look at the web from the side. How thick is it? Well, it is the thickness of a silk thread, just a few thousandths of a millimetre thick! The insect prey must be brought to a halt before that desperately thin barrier is breached. This is achieved by the use of a combination of specialized materials.

The points of attachment of web threads to branches and leaves, or to each other are made from the secretion of the pyriform gland. This attachment material contains a wide variety of amino acids, none dominating. The threads that form the frame of the web are the product of the major ampullate gland; the radii are the product of the minor ampullate gland. The amino acid composition of both these threads is similar and distinctive. Glycine and alanine are the main amino acids in both threads, and together form nearly 80 per cent of the composition of radial threads. Both frame and radial threads are obviously made of highly specialized polypeptides.

The capture spiral, product of the flagelliform gland, again has glycine as its main component, but alanine is much reduced in comparison with radial threads. Instead, proline is its second most common amino acid (Figure 6.3). This is an indication that the mechanical properties of the capture spiral thread are different from those of radii and frame threads, but that the threads collaborate in some way to arrest flying prey. The job of restraining the prey once it has been brought to a halt is assigned to the secretion that coats the capture spiral. It comes from yet another gland, the aggregate gland. The secretion of this gland has several components, but collectively they work to stick the web to the prey. The key component for this is glycoprotein, a typical component of mucus. The capture principle in the orb web of *Araneus* is based on mucus.

This glycoprotein is initially laid down as a continuous coating of the capture spiral, but then naturally aggregates into droplets. When a fly becomes stuck to the web, it is being held by dozens of sticky glycoprotein droplets. But there is a problem with this sticky droplet capture system. If the droplets dry out, they lose their stickiness. To overcome this, the aggregate gland secretion also includes compounds that are *hygroscopic*, that is, which can attract water out of the atmosphere. The coating of the capture spiral is normally about 80 per cent water. On dewy mornings in late summer it is the sticky droplets bloated with water that create those drooping, bejewelled webs.

So, that is the biochemistry of the different web components and the glands from which they come. Now it is time to look at the mechanical properties of all the web threads: frame, radii and capture spiral. To fully appreciate that, it is best to start at the molecular level and work up.

I've said that the amino acids of the web threads have more significance as engineering than as chemistry. This is seen in the arrangement of the glycine (lets now call it G) and alanine (A) that so dominate the composition of the radial web threads. In the bundles of long chain polypeptide molecules which make up a radial thread, these two amino acids repeatedly occur as alternating sequences of GAGAGA, although they may also occur linked to other amino acids. These GA repeats have a special mechanical significance which I will try to explain by means of a pasta analogy.

Imagine first that a single long-chain polypeptide is a piece of spaghetti. At intervals along the chain there will be GA repeats. They, you will remember, have the smallest possible side chains of all amino acids. In spaghetti terms, this means that where GA repeats occur, the spaghetti thread is fine and smooth, allowing strands of spaghetti to be laid side by side with GA repeats aligned. Chemical bonds form between the neighbouring strands, so that where several polypeptides are aligned in this way the effect is to convert parallel strands of spaghetti into a sheet of lasagne. Now imagine folding the sheet of lasagne repeatedly to the left, then the right, to form a tightly pleated sheet. This creates a compact three-dimensional structure of regularly arranged glycine and alanine building blocks, in fact a crystal. Where stretches of polypeptides do not have GA repeats, they do not align with neighbours to form crystals and remain as an irregular spaghetti tangle.

The radial threads of the *Araneus* orb web can therefore be said to have two sorts of zone: irregular and crystalline. The former are loosely arranged polypeptide chains in which are embedded numerous crystals of the latter. In web radii of *Araneus diadematus* the ratio of these two domains has been estimated at 68 per cent crystalline to 32 per cent amorphous.

The silk of web frame threads is a bit different. As in radial threads, about 40 per cent of the amino acid is glycine; the alanine component is about 18 per cent, which is a little higher than the amino acid proline. There are crystalline regions in frame thread but they seem to be largely made up of alanine repeats (AAAA), aligning with one another in neighbouring polypeptide chains to form pleated sheets. Glycine is, however, found in a variety of repeat sequences in combination with proline (P), for example the five peptide sequence GPGGX, where X can be one of a number of amino acids. Proline, unlike glycine and alanine, has a large side chain (Figure 6.3) and so cannot be compacted into a crystal, but it is thought that these five peptide sequences may adopt a spiral molecular conformation.

The main constituent amino acids of the capture thread are glycine (45 per cent), and proline (20 per cent). In this thread, alanine is the third most common amino acid, but at less than 10 per cent. The capture thread has neither GA nor AA repeats and therefore totally lacks crystalline regions. However, it is rich in the GPGGX repeats that are thought to adopt a spiral form.

That is the molecular background of the orb web: three different threads with different molecular architecture. This is what underlies their distinctive mechanical properties, which together perform the task of bringing fast flying prey to a halt.

You will almost certainly have heard it said that the strength of silk is comparable to that of high tensile steel. That is true, as far as it goes. However, it also gives a totally misleading impression of what spider's web silks are like. High tensile steel has less than 1 per cent extensibility before it breaks. It is very stiff. That's the whole point of using it in suspension bridge cables—that stiffness is what gives you a steady ride over the bridge in your SUV. The strength of *Araneus* frame or radius silk, that is, the maximum load that can be borne for a given cross-section, may be comparable to that of steel, but the maximum extension of frame thread before failure is a substantial 27 per cent and, for radial threads, 40 per cent. You would need to have bought your SUV for adventure rather than vanity to enjoy the bouncy ride you would get crossing a bridge suspended from these materials. I say 'bouncy' because the extension of these silk threads is elastic. Once the load of your SUV has been taken off the bridge, these imaginary silk suspensions will return to their former length.

Tensile steel is strong and stiff. The frame threads and radii of an *Araneus* orb web are strong, but they are not stiff. Suspension bridges need to give vehicles a steady ride. Frame threads and radii have a totally different job, to absorb the energy of impact of fast-moving prey.

The elasticity of spider silk thread is, however, a potential problem. As a rubber band is stretched, it stores energy. If I wrap one round the end of my thumb, pull back and let go, the elasticity of the material returns it rapidly to its original length, releasing the stored energy, which causes it to speed across the room. If spider silk behaved like that, it would stretch, storing the kinetic energy of the flying insect, bringing it to a halt. But it would then contract again, potentially throwing the fly off the web, back in the direction from which it came. Web silk, on stretching, needs to dissipate energy rather than store it. This it can do because it possesses the property of *hysteresis.*

I am going to explain hysteresis with the help of a graph. If you feel your eyelids getting heavy at the mention of such things, then you might be tempted to skip a couple of paragraphs. I urge you not to. How spider's webs overcome the problem of prey bouncing off the web is so elegant that it's well worth appreciating how it is done.

If I apply a stretching force to my rubber band, it gets longer. If I increase the force, it gets longer still. The graph (Figure 6.4a) shows

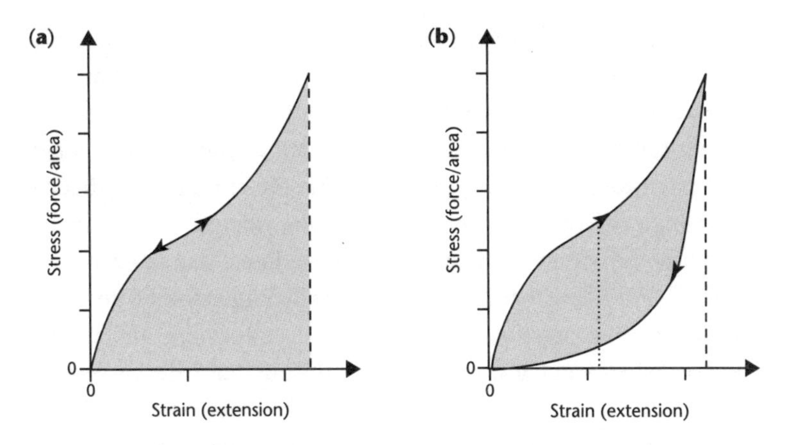

Figure 6.4. Silk absorbs the energy of a flying insect. (a) Curve of elastic extension and relaxation. The shaded area indicates the energy stored at full extension. (b) Curves of extension and relaxation for a spider's web thread. The dotted line shows where the thread is half extended. The shaded area between the curves shows the amount of energy dissipated through hysteresis.

a curve of force plotted against extension for my elastic band. The steep rise at the start shows that I initially get rather little extension for my increase in force. The curve then becomes less steep as I get rather more extension for not much more force. The curve then gets steeper again till eventually the force is so high that the rubber band breaks. The amount of energy released at this point is described by the shaded area under the curve. I'll come back to that in a moment.

Supposing I do not want to break my rubber band, and start to reduce the force I am applying just before reaching the breaking point. What will the graph show as the elastic returns to its original length? If it is perfectly elastic, it will contract along the same line in relation to force as it did during extension. None of the three silk thread types in the *Araneus* orb web behave like this, because of hysteresis. The graph (Figure 6.4b) shows what actually happens.

This figure shows the curve of force against extension for, let's say, a web frame thread. For convenience I have shown it taking exactly the same path during extension as the rubber band. Supposing I now relax the thread, allowing its length to return to half the maximum. Something odd has happened. The force now required to maintain this extension is much lower than it was during thread extension. Plotting the whole of the return curve back to the original length (zero extension) encloses an area between the curves. This shaded area compared with the total area under the extension curve, is the amount of the energy of impact of the flying insect that has now disappeared. For frame thread silk this is a really impressive 65 per cent.

Of course that missing energy has gone somewhere, and understanding where explains why the web threads have the molecular architecture they do. There is still some uncertainty about exactly what is happening at the molecular level but my pasta model gives a partial explanation. The complex molecular architecture of the frame and radius threads resists extension as the different components are forced to move past each other. Initially polypeptide chains of the amorphous regions move past one another as they straighten, crystalline regions are pulled into new alignments, and spiral molecular sections are perhaps extended. These rearrangements cause some energy to be converted into heat. When the tension on the thread is released and it returns to its original length, 65 per cent of that energy of movement of the fly has not been stored in the thread but dissipated as heat, greatly reducing the chances that the insect will be thrown back off the web.

The crystalline regions of the frame and radial threads have additional consequences. They probably help prevent cracks developing in the threads, making them tougher, and probably also make the threads stiffer. Cracks can propagate easily through a human manufactured mud brick but these can be made more resistant to cracks by the incorporation of animal dung, which is fibrous. When a crack being propagated through the mud meets a fibre across its path, the force concentrated at the apex of the crack becomes spread laterally and so dispersed, bringing the crack to a halt. The crystalline regions of spider's web frame and radius threads probably help stop cracks from passing through them.

Evidence that the crystals in the frame and radius threads increase stiffness comes from comparing them with the capture spiral. You will remember that the capture spiral is low in alanine and so lacks crystalline regions. It also has an astonishing 270 per cent maximum extensibility, i.e., it has exceptionally low stiffness. However, it does have 65 per cent hysteresis, allowing it, along with the frame and radial threads, to contribute to absorbing the energy of impact of a substantial flying insect. It is then the job of the special secretion laid on the capture spiral to restrain the immobilized prey.

You might think that was a full explanation of how the orb web works, but it isn't. There is yet another important component to the halting of the moving prey, one that operates not at the level of the web's different components, but of the web as a whole. It is *aerodynamic damping*.

When a web is hit by a flying insect, a wave of disturbance passes across its surface. This ensures that thread extension occurs by tiny amounts over a wide area. If you took a side view of the web shortly after the impact, you would see how the web was bulging out as the threads are stretched. But this means that, as a result of the impact, the whole web surface is being dragged through the air. The air resists the movement of the threads, causing further energy of movement to be converted into heat. This is aerodynamic damping.

You may imagine, like me, that the effect of aerodynamic damping is minimal, but it may be at least as important as hysteresis. It has been studied in the laboratory by striking an isolated web radius with the force of a flying insect and recording the strength and duration of its vibration, as you might for a plucked guitar string. If the radial thread is left attached to the rest of the web but struck in the same way as before, its vibration fades away and dies much more rapidly. This is because it has to pull not only itself but other web threads back and forth through the reluctant air.

The *Araneus* orb web is highly evolved at all levels, from the amino acid sequences and polypeptide chain architecture right up to the level of the whole web, to stop flying insects and prevent their escape.

These properties, we can now clearly appreciate, come not from cleverness in the construction behaviour but cleverness in the materials.

Throughout my description of the orb web I have referred to *Araneus* or specifically to *Araneus diadematus*. Different species of spider have different sorts of webs. I have already described specialized spider traps for catching ants and for catching pond skaters, but now I want to talk simply about specialized orb webs, ones for catching different sorts of flying insects.

There are quite a lot of species that build orb webs. These therefore superficially look much the same but in certain respects are quite distinctive and specialized. For example, although *Araneus* and a number of orb web builders use sticky droplets to prevent prey from escaping, another group of spiders, which includes orb web building species, use a quite different capture principle, something akin to micro-miniaturized barbed wire. In these species, the coating of the capture spiral, instead of being a sticky secretion of the aggregate gland, is a tangle of exceedingly fine dry threads produced by a structure close to their spinnerets called the *cribellum*. The cribellum is essentially a plate covered with a mass of tiny holes through each of which a very fine thread is extruded. This mass of threads is drawn off and 'backcombed' with a special structure on one pair of the spider's legs, then laid upon the core of the capture thread. Spiders producing this kind of thread are said to be *cribellate* spiders, because they possess a cribellum. Sticky droplet-producing species are therefore described as *ecribellate* (without cribellum).

There is a certain advantage to cribellar silk over sticky droplets: it is already dry, so it does not lose adhesiveness by drying out. On impact a small flying insect, which might be covered with numerous hairs or spines, finds itself ensnared at various points by the fluffy tangle of threads coating the capture spiral. Another adhesive principle may also be involved because, in addition, these threads seem to adhere to shiny surfaces such as the backs of ladybird beetles. We know less about the overall operation of cribellate than of ecribellate orb webs, but enough to know that they can be rather different. In the orb web of the cribellate species *Uloborus glomosus*, for example, the tensile strength of the threads is greater than in the cribellate *Araneus*, but they are also much stiffer, that is, breaking with less extension.

Consider for a moment, the orb web from the point of view of the prey. Flying insects have good eyes. This stops them from bumping into things. One of the things that they want to avoid bumping into is spider's webs. Over millions of years, spiders and their insect prey have been engaged in an arms race, with more and more sophisticated traps needed to catch more and more evasive prey. And just how many millions of years might that be? We have a 4mm length of spider's-web capture thread, complete with thirty-eight distinct sticky droplets, that is dated at between 127 and 132 million years old.[7] Well, strictly, it is a fossil of the thread, and it was recently spotted in a piece of amber collected in the Lebanon. We know from the fossil remains of the spiders themselves that they have been spinning silk for at least 400 million years, but this fossil capture thread is direct evidence of trap making.

It is this long arms race between spiders and their prey that has resulted in spider's web specialization, each design tailored to trap a particular type of prey. Insect vision has had a marked effect on orb web design and also its ecology. This is nicely shown in a comparison of two ecribellate orb web builders, *Mangora pia* and *Theridiosoma globusum*. The webs of *Mangora pia* have strong threads and a high-density mesh. The web threads of *Theridiosoma globusum* are finer and their density lower. Laboratory observations on the flight paths of fruit flies show that, even in a well-lit situation, they have difficulty in detecting the webs of *Theridiosoma* but begin to avoid *Mangora* webs from 7.0cm distant. Not surprisingly, therefore, the webs of *Mangora* are found in sites of dim illumination, and those of *Theridiosoma* in better lit places.

A possible habitat choice for spiders with more conspicuous webs is not simply dimly lit sites but total darkness. The Australian orb web spider *Eriophora transmarina* is nocturnal, while its compatriot *Nephila plumipes* is day active. The web of *Eriophora* has a larger area and has heavier sticky droplets than that of *Nephila*. It spins a new web every night, taking it down and eating it every morning to recycle the protein. *Nephila* webs by contrast are built to last several days. *Eriophora* could be characterized as a high spender in terms of its web building, but it is a high earner too; its capture rate is around twice that of the day active *Nephila*.

Another of the countermeasures evolved by flying insects to avoid being caught in spider's webs (that is, in addition to having good vision) is that of detachable body scales. These allow the insect to peal itself off the web after becoming stuck. This is the special adaptation of butterflies and moths. Countering this has led to the evolution of some extreme web designs, best exemplified by webs designed to catch night-flying moths, where the constraint of web visibility is relaxed. I want to illustrate two such designs, both making use of sticky droplets: a greatly elongated orb web, and a vestigial web.

Think of a moth striking a conventional orb web somewhere in the middle. It immediately starts to struggle, peeling off the wing scales and body hairs that hold it to the web. It begins to fall—not fall free—but fall down the web. It is still stuck, but by a different part of its body. It keeps struggling, eventually falling off the bottom of the web and escaping, leaving behind it a vertical smear of scales.

Had the web been longer, more like a broad ladder than an orb, then the moth might have run out of scales and become firmly stuck before it fell off. This very web design has evolved at least twice, both by modification of the orb web. In one of the species, found in New Guinea, it is the segment of the orb below the hub that has been greatly extended; in the other, found in the United States, it is the segment above the hub that is extended. In both, the vertical threads are greatly elongated radii, while the capture thread with its sticky droplets forms the rungs. These evolutionary changes are of course ones of behaviour, and not obviously ones of material. But they are simple changes to relatively simple behaviour. In laying down the capture spiral, orb web spiders not infrequently turn and go in the opposite direction rather than spiral continuously in one direction. This is most often seen at the edge of the web. The pattern of the rungs of a ladder in the ladder web is created by a simple repeated reversal of the thread direction in the extended web segment.

Mastophora hutchinsoni is also a specialist in catching moths, but its solution is an extremely reduced web. It has just one large sticky droplet at the end of one thread. To be precise, it does first produce a simple platform of silk threads to which the capture thread is attached, but it captures its prey by hurling the sticky droplet at passing moths. It is a so-called 'bolas' spider, named after the weights on the ends of interconnected cords thrown by South American gauchos to entangle the legs of running cattle but, archaeology shows, used by indigenous people pre-Spanish conquest to capture guanaco or the flightless rhea.

It might reasonably be expected that the chances of a moth, any moth, coming within striking range of a bolas spider would be vanishingly small, yet 90 per cent of the spider's prey consists of the males of only two moth species, the bristly cutworm (*Lacinopolia renigera*) and the smoky tetanolita (*Tetanolia mynesalis*). The spider can survive on this diet because there is an additional component to the trap—a bait. This bait is a cocktail of volatile chemicals that mimic the sex attractant signals released by unmated females of these moths. The unfortunate male moths are lured within range of the spider's trap by the prospect of sex.

Actually, it is a bit more complicated than that, and illustrates again the sophistication of self-secreted materials. The sex attractants released by females of both moth species are a blend of organic molecules, but they share no components in common. The spider has to produce two sets of attractant molecules. However, earlier in the night it catches mainly male bristly cutworm moths, but after 11 p.m. it starts to capture mainly smoky tetanolita males. Experiments show that this is mainly due to the moths themselves having different flight times, but that later in the night the spider also reduces the release of bristly cutworm attractant.[8]

So here is a baited trap built by a spider. The bait is not physically part of the trap, but it does draw prey to the trap and again it is a specialized self-secreted material. However, this is not the only spider to bait a trap. There are others where the bait is part of the web. Those spiders are in fact orb web builders.

It is a feature of the silk of the large primitive spiders, like the tarantulas, none of which produce webs to catch flying insects, that it reflects significant amounts of ultraviolet (UV) light, a part of the spectrum where insect eyes are particularly sensitive. It is significant that the silk of ecribellate orb web builders generally has a lower ultraviolet signature than the silk of the more primitive spiders, and that this is particularly evident in silk of webs that are typical of well-lit locations. The orb web threads of the spider *Argiope argentata* reflect low levels of ultraviolet light. Why, in that case, does this spider have emblazoned on its web, like the St Andrew's cross of the Scottish flag, two broad diagonal stripes of conspicuously white, UV-reflecting silk—silk produced by aciniform and piriform glands, material otherwise used to wrap prey?

Initially the strongest candidate hypothesis to explain this, and similar silk patterns in the webs of some other species, seemed to be that the silk gives additional strength to the web in windy sites. In acknowledgement of this they began to be known as *stabilimenta*. But the facts did not readily fit the hypothesis, and now such web devices, which seem to have evolved several times independently in the spiders, are just referred to as 'decorations'.[9] In *Argiope argentata*, decorations are more likely to occur in webs located in sheltered, poorly lit sites than in open, well-lit ones, the reverse of what the web-strengthening hypothesis predicts. This has led to alternative hypotheses: that the decorations distract predators, or alert birds which might otherwise accidentally damage the web. However, the sensitivity of insect vision to UV light has led to the suggestion that these patterns of silk are signals designed to attract flying insects.

Research and debate about web decorations are very active, but some good observational and experimental studies have tended to support the prey attraction viewpoint. For example, where the web decoration is in the form of an incomplete cross, prey impact tends to be in the web sector that has the UV-reflecting feature. Also, webs of *Argiope argentata* are more likely to include decorations when stingless bees, important prey for this spider, are more abundant. These are bees that are often attracted to flowers by the strong UV patterns they display.

Species-specific web designs are evidence of genetically determined inflexible building behaviour. However, there is growing evidence that some species of spider at least are able to modify their web-building behaviour as a result of experience. Spiders of the South American species *Parawixia bistriata* can build webs of two quite different types, each targeting a specific kind of flying insect.[10] Towards sunset they regularly build an orb web of about 8cm diameter with a tightly spaced capture spiral. This is designed to capture small flies. The other web is also an orb design and uses about the same amount of silk but it is much bigger, around 15cm in diameter, and with a much more widely spaced capture spiral. This web can be built at any time of day but characteristically after rain. It is a web designed to catch flying termites, those same plump, aspiring kings and queens that leave their home mounds after rain and that are trapped by the Azande to make porridge. Because the wings of these termites are long and their flight weak, the rather open mesh of the *Parawixia* termite web is quite sufficient to capture them.

Is this evidence of behavioural complexity in spiders? Almost certainly not. The behaviour that we observe could be generated by very simple rules of decision-making: 'If evening build small web; if after rain build large web.' Such a contingency could be inherited as a simple stimulus–response mechanism. There is no evidence here of a change in web building through learning; however, we now have at least some evidence that learning is involved in the web-building decisions of some spiders.

Orb webs, regardless of the species that build them, generally have a larger area below the hub than above it. This has been regarded as an adaptation to more efficient prey capture, since it is easier for a spider to run down the web to reach prey than to run up it. However, an experiment was conducted on spiders of a species *Larinoides scleroptarius*, in which young spiders were fed flies directly, so they never had to run over the web to catch them. When adult they were found to build webs that had almost an equal area above and below the hub, evidence that web proportions are influenced by experience. Furthermore, another group of these spiders raised on prey inserted in the upper part of the web, when adult, built webs with the sector above the hub enlarged.[11]

Clearly, there is a measure of sophistication in the web-building behaviour of spiders. I anticipate that in the next few years we will see more evidence of this. However, there will still remain, I am sure, a huge gap in the behavioural flexibility shown in the web building by spiders and in trap building by humans. Spiders will remain prime exemplars of the simple behaviour/clever materials route to become trap builders.

There is still a big problem that remains unexplained. Why are there no vertebrate trap builders other than the humans? It does seem extraordinary that until the very recent advent of human trap making, there was not a single vertebrate trap maker. Why are there no fish, bird or mammal trap builders of any kind, either taking our route, or taking the spider route?

It is useful to look again at the chordate *Oikopleura*, enclosed in its mucus house equipped with delicate filter nets (Figure 3.4). It has taken the typically invertebrate path, rather than that of its close relatives the vertebrates. Could not some true vertebrates have taken this path? The vertebrates have potential secretions for the making of traps, and indeed use some self-secreted materials for building. The kidney secretion of the male three-spined stickleback used for nest building has already been mentioned. The nest built by the male fifteen-spined stickleback (*Spinachia spinachia*) is of pieces of seaweed bound together by long strands of kidney secretion. Some coral reef fish of the wrasse and parrotfish families (Labridae and Scaridae) enclose themselves in a mucus cocoon to pass the night. Keratin, used to make horn, fingernails and hair, would seem to be a potential self-secreted, trap building material. The problem may be, at least in part, one of economics. Little swifts (*Apus affinis*) make nests entirely out of salivary mucus. It takes nearly two months for a pair to secrete a nest. It may be that, with the larger body sizes typical of vertebrates, the amount of self-secreted material required for a trap is prohibitively great, and so much more costly than alternative foraging methods of filter feeding, active hunting or ambush predation.

What about trap building with collected materials as opposed to self-secreted materials? In humans we see the combination of trap building from collected materials with the possession of a large brain, but is that a necessary association before trap building can evolve? I'm not sure that it is. Birds don't need to imagine nests, nor spiders imagine webs, in order to be able to build them. What birds lacked in that case was not the brains but the manipulative skill to be able to do it. Having just a beak may mean that the nest of a weaver is about the manipulative limit for working with collected materials. Making a trap would probably be equally complicated, and consider the return on investment compared with using beak or talons directly in food gathering. I suggest that no vertebrate was anatomically equipped to build an economical trap until the evolution of the hand.

It so happened that, by the time that vertebrates evolved skilful hands, they had also evolved large brains. The creatures that crossed those two thresholds were the humans. Was there a link between these two, the skilful hand and the creative brain? That will get further examination in the next chapter. Now we are certainly at a stage in the evolution of our brains where we are able to conjure up a multiplicity of trapping ideas. No other species has the brain capacity for such inventiveness, nor the technological mastery to translate them into substance. Humans will keep on designing new traps for capturing previously undreamed of prey, the dust of comets for example.

7

The Magic of the Tool Users

Imagine me facing a lecture theatre full of students, pointing out to them, as I did earlier in this book, that the height of a large termite mound is the equivalent of three times the height of our highest buildings. Having delivered this impressive statistic, I contrast this with our current realization that this is accomplished by a hoard of insects that communicate poorly with one another and almost certainly have no concept of what they are building. How different a termite mound is to this, I say, holding up a short stick, frayed at one end. This is a chimpanzee toothbrush. It was obtained from a group of captive chimpanzees that were using such tools to clean each other's teeth—chimpanzee dental hygienists!

As I challenged my students with this product of the creative chimpanzee mind, what I failed to notice was that through the half-open door of the lecture theatre had glided a Mercurian spacecraft. This had then landed on a vacant front-row seat, disgorging a party of tiny Mercurians led by the head of their Cosmobiology Research Unit. Interrupting me in a clear voice, she now said, 'What's so special about that? It's just a short stick, frayed at one end.'

The chimpanzee toothbrush story is true by the way.[1] What the Mercurian scientist said, although imagined, is also very apt. Look at the tools manufactured and used by chimpanzees: a crumpled leaf used as a sponge, a straight stem stripped of its leaves to form a flexible probe for pushing into termite mounds. As constructed objects, these are deeply unimpressive. They appear to require little manipulative skill and no assembly. Almost any kind of bird nest or caddis larval case should surely be held in higher regard in terms of its complexity? So what is it about a chimpanzee using a toothbrush that excites our interest? Well, it must be the chimpanzee's mind we admire, the creative thinking that we imagine was then translated into a novel, useful object: a tool. We believe that, unlike the caddis larva, the chimpanzee has used its brain to conceive of a new device and a clear plan of how to build it.

Let's take a look at a wild population of tool-using chimpanzees. A large adult sits holding a flattened stone, a hammer in fact, in its right hand. His manner is relaxed, lips closed in an almost-smile, eyes fixed on some distant, unimportant object. Glancing down, he picks up a rounded nut with the fingers of his left hand, placing it without fuss on the flat, level surface of a larger stone, an anvil. Quickly, the hammer delivers a sharp, measured blow, splitting the nut. The chimpanzee gathers the undamaged kernel with his left hand, popping it into his mouth. He looks up again at nothing in particular. To us as observers, this nut-cracking behaviour has transformed a forest ape into a personality of assurance and sagacity. We find it easy to make a link between the mind of the chimpanzee tool user and our own.

Tool making and tool use have been regarded for several decades as significant forces in human evolution. The influential anthropologist, Sherwood Washburn, wrote in 1959: 'It is my belief that the decrease in size of the anterior teeth and tripling in size of the brain came after man was a tool user, and as a result of new selection pressures coming in with the use of tools'.[2] Nowadays, our knowledge of the evolutionary transition from ape to man is much better documented from excavations of fossil bones and associated archaeology.[3] This is the history of our ancestry over about the last 6 million years, the history of the biological family Hominidae, which embraces us, our fossil ancestors, but not living ape species. Worked stone tools begin to appear in association with hominid bones in sites in East Africa dated about 2.6 million years ago. *Australopithecus garhi*, which lived in Ethiopia at this time, made stone tools, and (as judged by cut marks on excavated animal bones) used them to butcher meat. This was still a small creature by our standards (around 110–20 cm, that is less than 4ft), with a brain volume of about 400cu cm, close to that of the modern chimpanzee. It did not habitually walk upright and was probably still adept at climbing trees, but a creature similar to this was the ancestor to the first human species. One of these was *Homo habilis*, around 130cm tall and with a brain volume of about 600cc, which appeared about 2.5 million years ago.

The earliest *Homo* species differ from *Australopithecus* species in being capable of sustained upright walking. They also had less massive jaws and teeth. This, along with evidence from associated stone implements and animal bones, has led to the suggestion that these early humans not only used tools to hunt but also to process their animal and plant food, cutting it up, tenderizing it by pounding, and smashing bones to obtain the marrow. The result was a more nutritious diet, which in turn could have allowed the evolution of a larger brain. The connection between these two is not immediately obvious, but when you realize that our brains, while representing only about 2 per cent of our total body mass, consume 16 per cent of our daily energy intake, you can appreciate that big brains have substantial running costs. With bigger brains, it is argued, early humans could benefit from the virtuous circle again, improving their tools further, so improving their diet, and enlarging their brains yet again. We, *Homo sapiens*, now live in a world of complex technology, with highly processed food, and have brain volumes of around 1350cc. We have reason for wanting to know whether tool use has played a major part in raising us to the status of dominant world species through giving selective advantage to large brains.

There are alternatives to this 'tool use' hypothesis for the rapid increase in brain size of hominids over the past 3 million years. A particularly interesting possibility is that it was the advantage of being able to 'make friends and influence people' that drove growth in brain size. As our ancestors became more social, success depended more on social skills than individual strength, benefiting those with larger brains who could remember and understand social relationships. This would have led to greater social complexity, and given advantage to those with yet larger brains. This has been named the *Machiavellian hypothesis*[4] after the Florentine statesman and political philosopher, Niccolò Machiavelli, author of *The Prince*, a treatise on political manipulation that was published in 1532. Humans have other extraordinary attributes that may have contributed to our success, not least our phenomenal curiosity. What other species can you imagine saying, as the mountaineer George Mallory did in 1923 of his reason for wanting to climb Everest, 'because it is there'. This is no mere ironic footnote in the history of mountaineering; this is the attitude that led our species *Homo sapiens* out of Africa to colonize the world's land masses down to its smallest habitable islands in less than 100,000 years. Choosing mates on the creative brilliance of their brains may also have been important in the evolution of human brain size, and will be discussed in Chapter 8. So there are several hypotheses for the evolution of our large brains and consequent ecological supremacy—tool use, political skills, mate choice, and others besides. These explanations are not mutually exclusive, but the question here is the extent to which it was tool use.

An obvious problem in testing the connection between tool use and the evolution of human behaviour is that fossil evidence of this link is necessarily indirect. But we can directly study that connection in living non-human species that make or use tools. It is this that makes the study of tool use in all animals so interesting to us. Quite a variety of species show tool-related behaviour, if not tool making then tool using. However, this varied bunch are spread quite thinly across the animal kingdom. This rarity of tool users has strengthened the argument that there is something special about tools and that evolving tool related behaviour (making or using) is particularly difficult. From the perspective of human evolution, the most obvious difficulty would seem to be one of cognition. The animal needs to have some understanding of how to create a tool or, at the very least, how to use it to achieve a specific goal. But are animal tool users and tool makers particularly intelligent?

This is where I think we should pay attention to our immediate, almost visceral identification with the character of the nut-cracking chimpanzee. Even hardened scientists are not immune from this response; the danger of anthropomorphism when watching animal tool users is acute because they seem so human. We need to keep our sceptical guard up. Therc is another feature of tool use as a whole that also assists its ability to capture our imagination: the tools themselves. They are tangible and discrete; they appear to us as 'thinking made flesh'—objects we can collect and examine and contemplate. But, in that case, why isn't a bird nest also thinking made flesh and much more elaborate flesh at that? Whenever I ask that question—and I do, more and more—I get the answer that nests are just not the same, the implication being that nest building is essentially genetically programmed, and so does not involve intelligence or insight. But maybe we assume that nests are just not the same, without really having looked. Maybe we are not that interested in nests because we don't make nests, but my uncharitable side is inclined to say that tools have the advantage over nests in the public imagination because they are tangible, discrete and *not too complicated.* A bird nest is just a bit too much of a mess; a chimpanzee's straight stick, even frayed at one end, isn't.

So why are non-human tool users so rare? Well, we already have one hypothesis, which is that tool behaviour (using and making) requires a level of intelligence and manipulative skill that is hard to evolve (let's call it the '*tools, animal intelligence*' hypothesis). However,

I want to propose an alternative hypothesis for the rarity of non-human tool users. It is that tool use is rare because it is not very often useful (the 'tools are not often useful' hypothesis).

To support the 'tools, animal intelligence' hypothesis, we require evidence that the need for a brain capable of thinking and understanding has limited the evolution of tool-using species. To support the 'tools are not often useful' hypothesis we need to show that tool behaviour has had little impact on the ecology or evolution of tool users, and that it can evolve in animals with limited intelligence. If the latter hypothesis is supported in non-human species, we may have to admit that tool making and tool using demonstrate simply some rather limited examples of building behaviour. In other words, that the magic of the animal tool users is the magic of a conjuring trick, an illusion, and that their tools may have little to tell us about our own evolution.

Tool use is more fully studied in chimpanzees, both in the wild and in captive individuals, than in any other species. This is not simply because they are our nearest living relatives, but also because any chimpanzee troop that has been studied in the wild has been found to use tools, generally of a variety of different kinds. No other ape species—that is to say gorilla, bonobo, orang-utan or gibbon—uses tools in the wild to anything like that degree. Until recently, there was no clearly documented case of tool use in the wild by gorillas either. However, in 2004, a female gorilla was actually filmed using a stick to aid it in two different ways when wading upright into water up to her waist.[5] In the words of the scientific paper that described this discovery, she firstly 'seemed to use it to test the water depth or substrate stability: she grasped the stick firmly and repeatedly prodded the water in front of her'. After wading in, she then used it 'as a walking stick for postural support'.

This brief scientific paper received wide media coverage, but why? The way it was reported is revealing. The popular and respected

British science weekly, *New Scientist*, reported it under the headline: 'Gorilla Uses Tool to Plumb the Depths of Abstract Thinking'. Well, I like the 'plumb the depths' word play, but where did the 'abstract thinking' come from? According to the accompanying text in this account: 'the gorillas have understood that they can extend their sensory experience . . . by physically extending their bodies with an inanimate object'. Have they indeed! This looks to me like a straightforward example of an interpretation based on the reasoning: 'they behave like us, therefore they must think like us'.[6] If you feel that is a harsh judgement, then try this example of tool use, but this time by an insect, a predatory bug (*Salyarata variegata*) that feeds on termites.

The bug captures termites as they emerge on to the surface of their mound to extend or repair it. The bug creeps up to one of them, then grabs it with its front legs. It then stabs the termite with its stiletto mouthparts and sucks it dry. Then, instead of discarding the termite remains, the bug returns to the place where the other termites are still working, carrying the shrivelled husk of its first victim. The termites have a nest hygiene behaviour which includes removing the remains of their dead colleagues. One of the termites, detecting the termite remains being carried by the bug, moves forward to gather it up and dispose of it; suddenly it is the bug's second victim! The remains of the first termite have been used as a tool to capture the second.

Am I entitled to say that the bug 'understood that it could extend its capture range . . . by physically extending its body with an inanimate object'? And if not, then why not? Could I say the same thing of a spider sitting in the middle of an orb web? The original scientific paper on which the *New Scientist* article was based was more careful in its interpretation. Here, the use of the stick tool by the female gorilla is certainly regarded as an innovation rather than genetically programmed behaviour, but one that this female could have learned from another gorilla; nevertheless, the implication remains that some gorilla had been the inventor of this novelty. So was some termite-eating bug an inventor too?

The distant ancestors of this bug, we can fairly assume were not tool users. Their capture sequence was probably:

1. Approach.
2. Grab termite.
3. Suck termite dry.
4. Discard remains of termite.
5. Approach second termite.
6. Grab termite.

And so on.

Suppose, as seems likely to be the case, that this sequence is all essentially genetically determined and therefore inherited generation after generation. Imagine now that an individual inherits a mutation that causes it not to discard the remains of the first termite before seeking the second. It therefore produces the sequence 1, 2, 3, 5, 6, 4, and is thus a tool user, and incidentally a tool maker as well, since the live termite is transformed into a termite lure. Certainly there is novelty here, but the novelty is the product of genetic mutation. In case that seems a little improbable, here is a second insect tool use example to which we can attach the same kind of explanation. It is the digger wasp genus *Ammophila*.

Females in a few species of this wasp genus dig a vertical burrow into the dry ground, at the end of which is excavated a chamber where the wasp's larva will grow to adulthood. To achieve this, a female provisions the chamber with paralysed insect prey on which she lays an egg. She then refills the entrance shaft by dropping small stones into it. Finally, she uses a stone held in her jaws to hammer down the soil over the burrow entrance. The last stone is therefore a stone tool. If the wasp had dropped the last stone and just banged its head on the ground to firm the soil, then it would not be a tool user. As it happens, some *Ammophila* species do just this. Banging the ground with the head, it seems, was the ancestral behaviour of sealing the shaft. We can envisage tool use in *Ammophila* evolving from a simple genetically determined switch in that ancestral sequence, so that now the dropping of the final stone occurs after the action of hitting the ground with the head rather than before it.

So tool use is found in apes and insects. This has happened in both through an innovative process, but only in the former is there any possibility of innovation through the mind of an individual, and it is this fact in particular that seems to impress us. However, even where the possibility of a creative mind exists, we should be very careful in making that interpretation. Significant parts of the apparent innovation could still be determined genetically.

In the field of tool use, a brief check on definitions is advisable. A herring gull (*Larus argentatus*) breaking open the shell of a marine snail by dropping it on the paved promenade above the beach, or chimpanzee cleaning its ear out with its finger: are these examples of tool use? These and similar examples were considered in a book reviewing all then known examples of tool use, written by Benjamin Beck in 1980.[7] This book has provided the model for defining tool use ever since.

I won't go into the details of Beck's definition but, according to his criteria, the chimpanzee sticking a finger in its ear is not using a tool, because a tool must be an *unattached, environmental* object, which a finger clearly isn't. So, that also disqualifies shell smashing by the gull, because the promenade is an *attached* object, attached that is to its surroundings. However, an Egyptian vulture (*Neophron percnopterus*) picking up a stone and dropping it on to an ostrich egg is an example of a tool user because the stone is unattached, environmental and meets the additional criterion that the unattached object is *manipulated* in order to achieve a successful outcome.

I don't know whether you feel that there is more to be admired in the Egyptian vulture using the stone or the gull using the promenade, but now consider, alongside that, an invertebrate example. An orb web spider is not a tool user because the web, for all that the spider has to make it, is attached to its surroundings. The web becomes part of the environment and incapable, after its manufacture, of being manipulated. The ogre-faced spider (*Dinopis*), however, builds a much smaller web, so small in fact that it can hold it in its legs, capturing passing ants by pushing it down on them.[8] This, while constructed as a normal web, is then detached, making this spider both a tool user and a tool maker. What about a bolas spider (*Mastophora*—mentioned in Chapter 6, page 175), twirling its sticky droplet at the end of a silk thread. Is it a tool user and tool maker, or in fact neither? Well, that seems to depend on whether or not the thread is attached to the silk platform from which the spider hangs. You may remember with disappointment that the thread is attached to the platform; *Mastophora* is therefore just a web-building spider. These distinctions are beginning to seem all rather arbitrary but, more importantly, unhelpful. What we really want to know is to what extent construction and object manipulation require intelligence and understanding.

I read a paper published a couple of years ago on hunting techniques and tool use in North American badgers (*Taxidea taxus*) trying to catch ground squirrels. The badgers will often plug the ground squirrel burrow entrances to prevent their escape. The paper was careful to justify applying the term tool use to the burrow plugging behaviour, arguing that 'aimed movements of objects from up to one metre away from the burrow entrance' did qualify as tool use. So, a kind of badger and a kind of bug both belong to the tool user club, but is that telling us anything useful about human evolution or even about the cognitive powers of non-human tool users?

Let's get back to the tool using apes. We share, as we are fond of telling ourselves, about 98 per cent of our DNA sequence in common with chimpanzees. However, there is no clear correlation between the frequency and variety of tool use among the apes and their relatedness to us. All wild chimpanzee groups have been observed to use tools at least occasionally, but bonobos (our next closest relative) rarely use tools in the wild. Gorillas only very occasionally use tools in the wild, yet orang-utans, which are less closely related to us than gorillas, readily use tools in captivity and in some locations are tool users in the wild. There is also no clear correlation between brain size in apes and prevalence of tool use. Having made an adjustment for body size, the chimpanzee, among the apes, does indeed have the largest relative brain size but then comes the gibbon, the least among the apes in terms of tool use, then the orang-utan, and then the gorilla. However, in spite of these poor correlations, the chimpanzee remains our nearest living relative and a habitual tool user. We do need to study it.

If we are looking for signs of intelligence in chimpanzee tool behaviour, what should we be looking for? Well, any behaviour in any animal is a product of the interaction between genetic factors inherited from the parents, and experience. If tool-related behaviour generally requires complex mental abilities, then we might anticipate a long learning process, not simply through trial and error but also from copying the example of others. We might also expect to see evidence of understanding or cognition, for example of spatial relationships, or of the link between cause and effect; we might hope to see examples of innovation or invention. There are also two aspects of human behaviour that, although evidence of our sophistication, are at least worth looking for in our nearest relatives: these are education and culture. We facilitate learning in our children by sending them to school. Do chimpanzees teach their young to use tools? Much of what we learn, whether through personal experience or being taught, is functional and adaptive, but some of it just reflects the culture in which we were raised. Language is our most obvious example of this, but it can be seen in regional differences in architecture and decoration. Some objects or buildings just look Chinese or Russian. Where these differences have no clear adaptive significance but are just expressions of 'the way we do things around here', they are cultural differences. They are expressions of human creativity. Are there cultural differences in chimpanzee tool making or tool use?

Finally, having listed the features of tool behaviour that might persuade us that the animals showing it are intelligent (brain size, learning process, cultural transmission), let us also not forget the clear and important message from invertebrate examples, such as the hammer of the digger wasp *Ammophila* and the portable web of the ogre-faced spider *Dinopis*, that tool use and even tool making can be achieved by essentially genetically determined behaviour alone.

Jane Goodall's pioneering study of wild chimpanzees in Uganda in the 1960s revealed their preparation of a variety of tools, used in a number of contexts.[9] Probably the most impressive of all these was the use of flexible plant stems to catch termites from inside their mounds. The principle of this termite 'fishing' is that damaging a bit of the surface of a termite mound not only allows a narrow plant stem to be inserted into it, but also alerts the large, soldier termites to defend the nest. This they do by grabbing at and locking their jaws on to the plant stem when it is pushed in through the hole by the chimpanzee. A skilled, experienced chimpanzee can then carefully withdraw the stem and eat the attached termites—a useful protein supplement to its diet.

Jane Goodall's discoveries provoked a burst of field studies to look for tool use in wild populations of primates. For chimpanzees, new evidence quickly came to light: new types of tool and different tools to do similar things in different places. Among these were more examples of tools used to feed on social insects, not simply termites but also on the highly aggressive nomadic colonies of driver ants (*Dorylus* species), by a technique referred to as 'ant dipping'. The ant dipping tool is typically a straight, stiff stick, about 60cm long and 1cm wide. To make it involves breaking off a straight branch, removing all the side branches or leaves, and sometimes even peeling away the bark to create a simple smooth wand.

Ant dipping begins with the insertion of the narrower end of the wand into the subterranean nests of the driver ants (*Dorylus*). The chimpanzee grasps, one-handed, the other end. The aggressive ants stream up the wand in defence of their nest but, when they have got about three-quarters of the way up, the chimpanzee withdraws the tool from the nest, points the narrow end at its mouth and, with a smooth, sweep of the free hand, bundles a couple of hundred ants into its mouth and chews rapidly.

As discussed earlier, stiff stick tools are also employed by chimpanzees to feed on termites. At one site in the Congo, a sturdy stick is used as a digging tool to breach the termite mound wall, then a finer, brush-ended probe is used to pick up the termites and deliver them to the mouth—two tools used in combination, a *tool set* in fact. An even more impressive example of a tool set is the report of a female chimpanzee extracting honey from the nest of a colony of stingless bees (these often have very painful bites by the way), using a combination of four tools in sequence: first a stout chisel, then a fine chisel to break through the wall of the nest, then a fine bodkin-like tool to pierce into the honey storage area, and finally a dip stick repeatedly inserted into the nest to be withdrawn dripping with honey.

Also discussed earlier, the use of hammer and anvil to crack palm nuts by chimpanzees in Guinea, West Africa, requires that a set of tools be brought into a particular spatial relationship with the nut in order that they can be successfully used. The anvil must be hard and level; the nut must be placed on the anvil and then struck with something hard but in a measured way, so as to break the shell without crushing the kernel. An instance has been reported in which a chimpanzee used a stone placed under the anvil to make it more level; in other words, using a tool to assist the operation of another tool.

Observations from the field suggest that, before young chimpanzees become efficient nut crackers, an extended period of observational learning and practice is involved. From an early age, young chimpanzees take great interest in and interact with experienced nut-cracking adults. They also imitate elements of the sequence: hitting the ground with a stone or a nut with the hand. In one engaging sequence I have seen on film, a juvenile chimpanzee holds a nut precariously on an uneven anvil with the toes of one foot, which are only withdrawn just as the handheld hammer delivers its ineffectual blow. No chimpanzee under three years of age has been seen to use the hammer and anvil technique successfully, and some individuals, in a ten-year study of the group, have never achieved it.

Evidence of teaching of nut-cracking technique is disputed. There have been claims of mother chimpanzees helping to educate their children by leaving a hammer on an anvil, or a hammer, anvil and nut all together, but the cautious interpretation is that this is accidental. Certainly if teaching is occurring, its importance is marginal.

Nut cracking is known from another site in West Africa but here other hammer and anvil techniques are seen, for example, wooden hammer and stone anvil, or wood for both hammer and anvil. So, is this a cultural difference (an example of 'this is the way we do things around here') or an ecological one (an example of 'this is the best or only way to do things around here')? Distinguishing between these two is difficult because it is hard to be sure that every possible ecological explanation has been eliminated. Regional differences in the length of wands used in ant dipping on closer examination in the field did suggest an ecological, not a cultural explanation. Longer wands were used to capture driver ant (*Dorylus*) species that give more painful bites.

Other evidence of variation in the use of tools by chimpanzees does, however, support the view that cultural differences do exist. A detailed consideration of regional differences in all behaviour seen in wild chimpanzees across Africa, conducted in 1999, concluded that thirty-nine different behaviour patterns, fifteen of them involving the use of tools, others including styles of grooming and courtship, exhibited differences between locations for which no ecological reason could be found.

These field studies on chimpanzees and other apes in their natural habitats have now revealed quite a lot about what they do, but not a lot about what they think. That evidence has come in particular from studies on captive apes using tests reminiscent of those carried out by child psychologists.

I don't want to cover the range of tests used to study chimpanzee cognition and the capacity of this species for abstract thought. Our subject here is animal tool use and its significance in the understanding of human evolution, so I want to look in particular at the

Figure 7.1. Trap-tube test: a chimpanzee successfully uses a tool to push a peanut out of a tube without it falling into a trap.

Fig. 4.1a (p. 110) from D. J. Povinelli, *Folk Physics for Apes*. By permission of Oxford University Press

performance of chimpanzees in two psychological tests involving tool use. That way we can explore whether chimpanzees think like us when using tools. In the first problem, the chimpanzee must use a stick to poke a food reward out of a transparent tube; in the second, use a rake to drag food towards itself.

If a chimpanzee is shown a peanut inside a transparent tube and provided with a long enough pole, it will readily learn to insert the tool into the tube to push the nut out of the far end. This tool use problem can also be solved by some other primates, for example South American capuchin monkeys.

In a particular experiment, four chimpanzees, having been trained to solve the stick in the tube problem, were presented with a variant of it in which there is a trap to catch the peanut positioned in the middle of the tube—the *trap-tube test* (Figure 7.1). The peanut is placed in the tube so that it is close to the trap into which it will fall if pushed in the wrong direction. The tricky part, as you can see, is that to push the peanut out safely, the stick must be inserted at the end furthest from the nut. Only one of the four chimpanzees tested was able to master this. But what did she (Megan) think she was doing? Did she understand the concept of avoiding the trap (that is, understand the causal relationship between herself, tool, trap and reward) or just learn to carry out a successful procedure?[10]

In an attempt to answer these questions, Megan was given a further test, in this case with the transparent tube rotated 180 degrees on its axis so that the trap pointed upwards and consequently no longer functioned as a trap. Megan, presented with the peanut in the tube in the same position along it relative to the trap, continued to use the same procedure in spite of the fact that it no longer mattered in which direction the nut was pushed. But, then, why shouldn't she? The pole was placed mid-way between the two ends of the tube and so could equally readily be taken to either end; why not just follow the previously successful pattern? However, even when the pole was placed at the end of the tube nearest to the food reward, Megan persisted in taking the pole to the other end to push the nut out, supporting the view that she is following a procedure, not understanding causal relationships.

The second tool use test involves the principle of a rake to gather food that is out of reach. A chimpanzee provided with a 'toothless' rake, of the sort used in casinos to gather gambling chips, quickly learns to use it to reach and gather peanuts or whatever, but what is its understanding of how the rake works?

In one experiment, seven chimpanzees were separately presented with two identical rakes, both with only two prongs, one at each end of the horizontal cross bar. The heads of the rakes were then placed beyond the reward, one with prongs down, and one on its back with the prongs in the air. The first is obviously ineffectual because the arch created by the two prongs passes right over the reward, but the horizontal bar of the rake on its back scrapes along the ground and so can be used successfully.

I say 'obviously', because it is obvious to us. The chimpanzees, however, performed no better than chance in their selection of the ineffective or effective rake. Just in case they had not noticed that the prongs of the rake were of a length that the reward would pass under the arch, a further test was done in which chimpanzees were given the same choice but using rakes with substantially longer prongs. The performance of the chimpanzees did not improve.

There are other tests of this kind using tools that have produced similar results in the sense that, by our reckoning, the performance of the chimpanzees has been surprisingly poor. But this is based on the reckoning that since they behave like us, they think like us. Certainly in these tests, they do not think like us. In particular, their understanding of causal relationships is poor compared with ours. That being the case, it is useful to compare the nature of tool use in chimpanzees with other mammals and with the most proficient of tool making and tool-using birds. In that way we can assess what they (non-human tool users), collectively think about the tools they are using, and whether that tells us anything about the role of tools in human evolution.

Nut cracking using a hammer and anvil, which because of its use of two complementary tools seems a particularly sophisticated form of tool use, is not confined to chimpanzees but is also shown by wild capuchin monkeys *Cebus libidinosus*, albeit in a somewhat simplified form.[11] As with chimpanzees, the occurrence of this behaviour coincides with appropriate ecology—ready access to hammers, anvils and nuts, and limited availability of other sources of food.

The monkeys at the site where this behaviour occurs don't use rock tools for anvils, instead taking advantage of areas of flat exposed rock, some dimpled apparently by the repeated smashing of nuts. Another notable difference between their behaviour and the nut cracking of chimpanzees is the relative size of the hammers they use. Capuchin monkeys are small, (about 3kg, compared with 40 to 70kg for a chimpanzee), so to crack the local palm nuts, a capuchin will pick up a rock that can regularly weigh 20–25 per cent of its body weight, stand upright and drop the rock from shoulder height on to the nut (Figure 7.2). The cognitive abilities of capuchin monkeys have also been looked at using some of the standard psychological tests, so it is interesting to note that in the tube test with the trap in the floor and

Figure 7.2. Tool use by a capuchin monkey: the monkey cracks a tough palm nut by dropping a large stone from an upright position.

Pete Oxford/Minden Pictures/FLPA

with the trap rotated so it is in the roof, capuchin monkeys perform in essentially identical manner to chimpanzees. They appear simply to happen upon an effective procedure then stick to it, but show no evidence that they understood what made it effective. I don't intend this as a patronizing judgement on the limitations of the minds of these monkeys or of chimpanzees, but rather an illustration of the effectiveness of a simpler methodology. Sticking to doing something in a certain way because we have learned that it works is, after all, a tactic that we habitually adopt ourselves.

Anecdotal evidence adds several other mammalian tool users to the list, but few that we know much more about. However, there is some evidence that tool use in a population of bottlenose dolphins (*Turscops* species) off the coast of Australia persists through social learning.[12] A few individuals in this one location regularly carry sponges held in their teeth that cover the tips of their noses. It is still not clear what these dolphins are doing, but it seems that they use the sponge tool to probe and sweep places on the seabed where not only prey species but venomous species like scorpion fish and sea snakes may also be concealed. The sponges are their equivalent of wearing protective gloves. Mitochondrial DNA analysis obtained from tiny biopsy samples has shown that these dolphins are a group of related females, strongly suggesting social transmission of the tool-using habit from mother to female offspring.

Tool use in birds seems to be a little more common than in mammals, even allowing that there are about twice the number of species in the former than the latter. A survey of the scientific literature published in 2002 came up with a list of 104 bird species showing undisputed or at least arguable instances of tool use.[13] Interestingly, the relative brain sizes of the species showing clear evidence of tool use were found to be significantly larger than species regarded as marginal tool users. However, whether that larger brain size is associated with tool behaviour rather than some other behaviour, or indeed indicative of greater intelligence, needs further evidence.

This same survey showed that tool use was particularly prevalent in the crow species, but is also sufficiently widespread that we can be confident that tool use has evolved several times independently in birds. However, the best evidence we have of the nature of tool making and tool use in birds comes from two species: the New Caledonian crow (*Corvus moneduloides*) and the woodpecker finch (*Cactospiza pallida*).

The New Caledonian crow, a species confined to New Caledonia and some neighbouring Pacific islands, is certainly not a marginal tool user. It not only uses but also makes tools and of more than one kind. In fact it became something of a celebrity when its tool making behaviour was first described in 1996 by Gavin Hunt who, with his co-workers at the University of Auckland, has been responsible for continued field studies on this species. What Hunt observed was the manufacture of two types of tool, both of which use the principle of a hook to extract insects and similar prey from crevices.[14] One kind of tool is a narrow flexible twig devoid of side branches or leaves, with a recurved spur or hook at the thicker end (Figure 7.3a). This tool is created by the bird biting a fine twig at its base, at the same time including a piece of the main stem above the point at which the twig is attached to form a hook. Held in the beak at the narrow end, this twig is used as a hooked feeding tool.

In the second type of tool made by New Caledonian crows, the hooks come ready made on the serrated edge of the leaf of a *Pandanus* palm. The bird creates the tool by cutting a narrow strip of leaf in a characteristic way. It first makes a short diagonal cut with its beak in from the edge of the palm leaf and then tears along the parallel leaf veins in the direction of the leaf tip. It repeats the cut-and-tear sequence perhaps another twice before defining the wide end of the tool with a new cut, followed by a tear that meets up with the first. This completes a 'stepped-cut' tool (Figure 7.3b). The wider end of the tool is then held in the beak so that the narrow end can be inserted into crevices and the recurved spines used to extract prey. The stepped-cut tool can be cut from either leaf margin, the essential

Figure 7.3. New Caledonian crow makes tools.

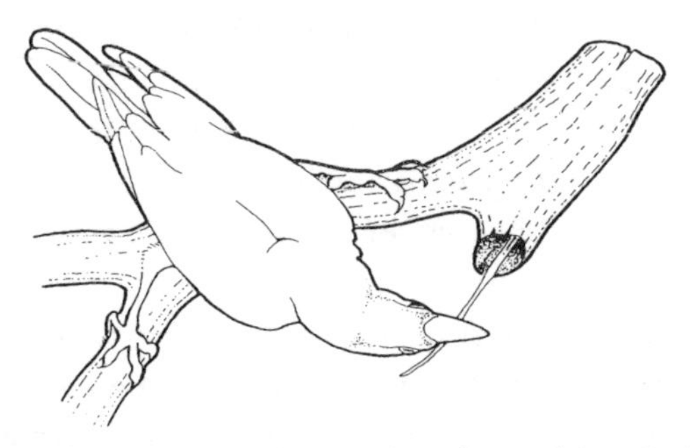

(a) A New Caledonian Crow probes a cavity using a stick tool with a hooked end that it has fashioned itself

Adapted by permission from Macmillan Publishers Ltd: Nature (1996). Gavin R. Hunt (1996), The manufacture and use of hook-tools by New Caledonian crows. *Nature* 379: 249–51

(b) 1. A stepped cut tool is made by cutting out a small part of the edge of a pandanus leaf (arrowed) 2. The tool is cut so that the spines on the leaf edge point towards the wider end of the tool; the fine end of the tool is used to extract insect prey from crevices

Adapted from Hunt, G. R. (2000). Human-like, population level specialization in the manufacture of pandanus tools by New Caledonian crows *Corvus moneduloides*. Proceedings of the Royal Society of London B, 267, 403–413 and from Hunt, G. R. and Gray, R. D. (2004). Direct observations of pandanus-tool manufacture and use by a New Caledonian crow (*Corvus moneduloides*). Animal Cognition 7, 114–120

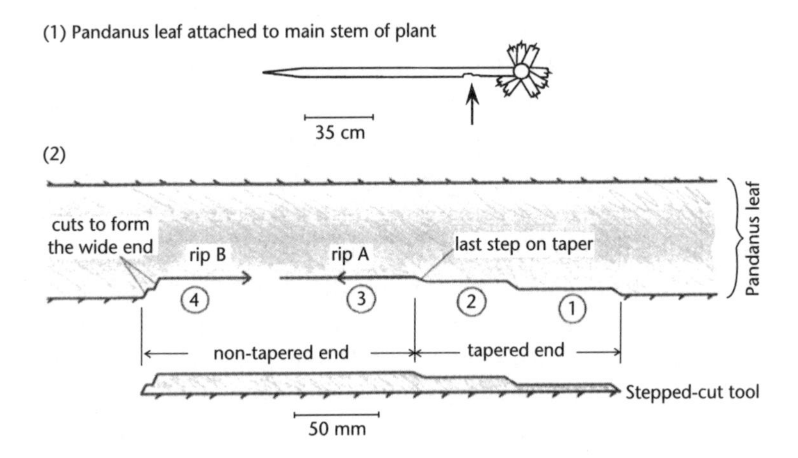

feature being that the hooks must be directed towards the wide end of the tool.

There are plenty of scientific reports and popular anecdotes supporting the view that crow species in general are intelligent and flexible in their behaviour. We should therefore take seriously the possibility that New Caledonian crows are, in their tool making and tool use, exhibiting behaviour that is comparable in its complexity to that shown by chimpanzees and that might tell us something about ourselves. But what exactly are the New Caledonian crows thinking?

Hunt's initial assessment was that the crows' behaviour represents something really quite special, showing, he suggested, features of tool making that 'only first appeared in the stone and bone tool-using cultures of early humans after the Lower Palaeolithic'. This equates the tools of the crows with the stone tools found in association with *Australopithecus* and the first *Homo* species.[15] This judgement is based on several features of the New Caledonian crow tool behaviour, in particular, that there are two quite distinct tool types, that they both show a high degree of standardization in their manufacture, and that the manufacture requires precision manipulation. In addition, Hunt emphasizes that the tools exhibit clear evidence of functional design, including the incorporation in both of the principle of a hook. His conclusion is that the crows show evidence of foresight and planning, a conclusion reached, as you see, on the basis of both manipulative skill and cognitive ability.

I would really like to agree with this, but find myself unable to do so. My problem is that whereas I am being asked to compare the performance of the crows with tool making in apes and ancestral humans, I find myself comparing it with any kind of building behaviour by both vertebrates and invertebrates. Against this background, the crow's achievement seems much less remarkable.

Take the issue of standardization: I was arguing in Chapter 3 of this book, that standardization of building units was the small-brained solution to creating orderly structures with a strong element of design. As to the complexity and manipulative skill involved in the creation of, for example, the stepped-cut tool, I mentioned in Chapter 3 the 'box girder' case built by the caddis larva *Lepidostoma hirtum* from somewhat rectangular leaf panels. I don't mean 'somewhat' in a vague sort of way, since each panel has a very particular shape: convex at the anterior edge, concave at the posterior edge, with the sides both showing a convexity greater than that of the anterior margin. These panels are fitted together so that, in each of the four sides, the convexity of the anterior edge of one panel fits into the concavity of the trailing edge of the panel in front. These joins between the panels of one side of the case are half a panel length out of phase with those in the neighbouring sides, and consequently in phase with those in the opposite side; as a result, no lines of weakness pass straight across the box-grinder: evidence, it seems to me, of standardization, manipulative skill and functional design.

I should certainly concede that the New Caledonian crow makes not one but two standard designs with different materials, each to a distinctive specification, but the argument of special manipulative skill required by the New Caledonian crow seems unconvincing. A New Caledonian crow does not, in fact, even need to cut all the sides of the stepped-cut tool. It must certainly make all the oblique cuts at each step and the final perpendicular cut to separate the completed tool from the *Pandanus* leaf. However, the perfectly parallel sides seen along the length of the tool are a property of the leaf itself, its parallel veins, reminding us again of what we established in Chapter 3, that the construction behaviour may be kept simple by exploiting properties of the building material.

Consider how special this example of manipulative skill is. A bird strips all the leaves off a living twig. It then snaps through the twig, not cleanly, but so as to tear the twig away from the branch with a long flexible tongue of bark projecting from its broken end. The bird then takes this twig to the tip of the branch of a tree and binds the soft tongue of bark round it. The tongue of bark dries, fastening the dangling twig to the branch. The bird then repeats this process over and over again until it has built a complete hanging chamber with a downward-directed entrance tube. The bird is a male red-headed weaver (*Anaplectes rubriceps*), and it is building a nest.

Some aspects of the field observations on New Caledonian crows cannot so readily be dismissed, in particular the variation in the design of the stepped-cut tool. The number of steps in *Pandanus* palm tools cut in the wild varies from one to four; in addition some tools are just the same width along the whole length with no steps, but these are of two types as well, wide and narrow. More interesting still is that the tool variants show distinctive regional distributions that are not obviously explicable by ecological differences. This could be interpreted as cultural differences maintained by social learning, as seen in chimpanzees, although it is too early to rule out the possibility that at least some of this regional variation is in fact genetically based.

It would suit me to be able to say at this point that nest building of a species of weaver bird, for example, shows regional variation and, even better, add that we know this to be culturally or genetically determined. Regrettably, I can't, but the reason is that no one has looked. Maybe such differences exist, maybe they don't. I am currently planning a trip to Africa to obtain material to test this. However, I can't escape the conclusion that the reason we know about regional differences in the stepped tools of New Caledonian crows, but not in the nests of weaver-birds, is that it is a lot easier to count the number of steps on a *Pandanus* leaf tool, than the number and geographical distribution of interlocking loops, half hitches, and slip knots in the nests of weaver-birds.

There is another piece of evidence from field studies on New Caledonian crows that suggests mental sophistication and which therefore needs to be considered—'handedness'. You will have heard of, possibly even know of, a victim of a stroke who lost the ability to speak due to damage to the left frontal lobe of the brain. This and other lateral asymmetries in the operation of the human brain are evidence of greater specialization in our brains compared with other mammals. Functional asymmetry is therefore regarded as being an important step in the evolution of the cognitive abilities that characterize modern humans. One piece of behavioural evidence of lateralization of brain function in humans is 'handedness' and dominance in the use of the right or left hand. The scars left on *Pandanus* leaves by New Caledonian crows after removal of stepped tools from the left or right leaf margins indicate that in the wild the crows exhibit handedness. Four wild-caught New Caledonian crows were seen to show 'handedness' in how they held a tool, two regularly placing the non-working end of the tool across their left cheek, the other two, the right cheek. Similar handedness is also shown by wild and captive chimpanzees.

Greater understanding of how New Caledonian crows think when making and using tools has come from the experimental psychology approach, in particular from a group at the University of Oxford group led by Alex Kacelnik. For example, in an experiment very similar to that already described for chimpanzees, two captive, wild caught New Caledonian crows were offered food (in this case meat) in a transparent tube. Presented with a rack of sticks of different length to push the piece of meat out of the tube, the crows readily used the tools provided. Generally they chose a stick that matched the distance from one end of the tube to the other, or chose the longest stick provided, demonstrating ability at least comparable with that of chimpanzees.

To test their understanding of the operation of hook tools, two wild-caught crows were presented with a choice of wire tools in order to lift a bucket containing a piece of meat from the bottom of a vertical, transparent well. The crows proved able to solve this task; when given a choice of a straight wire or one with a hook, they chose the latter. More than that, one of the crows (Betty), finding herself left with only the straight wire, bent the end to create a hook to retrieve the bucket![16] Isn't that an example of insight and invention? In order to test this, Betty was asked to make a hook tool, this time not with wire but a narrow, straight strip of aluminium sheet.[17] The logic of the experiment is that to make a hook tool now requires new behaviour, firstly because it is a different material from the wire, secondly because the rectangular section of the material only allows it to be bent in one plane. By her third trial, Betty had successfully made and used the new hooked tool.

Interpretation by the researchers of the results of this hook tool experiment is notably cautious. They note that Betty starts by treating the aluminium strip as if it is a wire, although soon learning that it has different properties. However, she also persists in trying to obtain the food with an unmodified metal strip at the start of a trial even after she has been successful in bending it in a previous one. They conclude that she is not showing insight, but neither is she simply following a procedure. She has some understanding, and that needs further study. An obvious problem in interpreting her behaviour is that, since she was caught in the wild, we have no idea of her previous experience, but now we are beginning to get results from aviary reared birds.

Two hand-reared New Caledonian crows were placed as juveniles together in an aviary that contained sticks of various shapes and sizes. They were also given regular demonstrations by their human foster parents, to which they paid close attention, of how to use twig tools to extract food from confined spaces. Both started to handle twigs and to obtain food with them. However, two other young crows were each placed alone in an aviary and given no tool using tutorials. Both of them began to use the stick tools, and just as fast as did the tutored two.[18] In fact, one of the solitary housed crows, on the first day of being presented with a *Pandanus* leaf, cut a parallel-sided hooked tool with a swift *cut-tear-cut* action, immediately took it to a crevice where food was often hidden and used it as a probe, later using it successfully as a hooked tool.

New Caledonian crows, it seems, have a strong inherited disposition to make and handle tools of a certain kind. Trial-and-error learning may be important in the wild; social learning may also occur, and it may be responsible for at least some of the observed regional differences. However, we may not need to invoke either to account for some of the basic elements of tool use, and indeed tool-making behaviour, that we see in this species. That, after all is the message of the invertebrate tool users. We should also not be surprised to see behaviour develop through complex learning linked to strong, genetically determined developmental guidelines. Such systems are well known in birdsong development for example. An analogous example from domesticated animals would be the behaviour of sheep dogs. That is the border collie type of sheep dog that performs at sheep dog trials.

What the audience see is the dog circling and crouching, cajoling five reluctant sheep into a small pen in response to a strange vocabulary of whistles and calls from an immobile human master—clear evidence that painstaking training has been involved. But this champion sheep dog comes from a long line of dogs bred specially for their aptitude in herding sheep. She was chosen carefully as a pup for the inbred behaviour (crouching, staring at the sheep and circling around them). These genetically based traits allowed her to be trained to become a champion.

I want now to move the spotlight from the New Caledonian crow to the other bird species that is almost as celebrated for its tool use. *Cactospiza pallida* comes from the Galapagos Islands and was first collected by Charles Darwin. Initially it was known by the almost perversely dull name of the *pallid finch* (still its scientific tag), although it shared the celebrity of the other 'Darwin finches' as being a classic example of the evolutionary process. However, early in the twentieth century it was discovered to use cactus spines and similar sharp probes held in the beak to remove insect prey from crevices. It was a tool user, deserving to be a celebrity in its own right. It was renamed, rather more glamorously, the 'woodpecker finch'. It is also a tool maker. It manufactures tools in the same way that a chimpanzee will manufacture a probe tool, not simply breaking a cactus spine or stick off the plant but shortening it or removing side projections.

So is social learning involved in the development of tool behaviour in the woodpecker finch? This was tested first by placing ten adult woodpecker finches that were unable to use tools in an aviary with some tool-using companions.[19] All remained unable to use tools. Next, a group of seven naïve young finches were housed with experienced tool-using individuals; all learned to use tools. However, before we jump to conclusions, a control group of young birds with no tutors also developed tool-using behaviour and with equal speed. Social learning, even if it does occur in this species in the wild, is not essential.

Inevitably, woodpecker finches have also been tested with the problem of the horizontal transparent tube containing a food reward. All five birds that were offered a rack of sticks of various lengths eventually used one as a tool to push the food out of the tube. However, there was no clear evidence that they tried to match the length of the stick to the task. Nevertheless, two adopted the successful tactic of: if at first you don't succeed, next time choose the longest stick on offer.

In a second test, a stick tool was offered which had cross-pieces at both ends to prevent it being inserted into the tube until at least one cross-piece was removed: a so-called *H-shaped tool*. Three of five woodpecker finches tested did solve the problem, but not in a manner that showed great insight. Two took fourteen trials of fifteen minutes, the third twenty-one trials. Even after achieving success, all three continued to make mistakes. One, for example, repeatedly removed the cross-piece from one end of the stick to create a *T-shaped tool*, then tried to insert the end with the remaining cross-piece into the tube.

Interestingly, chimpanzees tested with the H-shaped tool and the tube are about equally as unconvincing as woodpecker finches in demonstrating insight and understanding. Both species seem to depend largely upon persistence, with changes in tactics until something works.

Woodpecker finches have also been tested on the problem of using a stick tool to remove the food reward from the transparent tube with the trap, like the one used to test chimpanzees (Figure 7.1). Only one of six finches tested was able to solve this. When this female

(Rosa) was tested with the 'trap' now positioned in the roof of the tube and therefore ineffective, she behaved rather differently from the chimpanzee which, you will remember, still behaved to our way of thinking as if the food might fall into the trap. Rosa instead simply persisted in removing the food from one end of the tube, regardless of the position of the food relative to the 'trap' above it. So, maybe she understood that the trap was no longer able to operate or maybe she just followed a successful procedure.

I need to try to summarize the significance of animal tool making and tool use, but, first, a story that you may have heard concerning the behaviour of a dog:

> A man in a railway carriage looks up from behind his newspaper and sees opposite him another man playing chess with his dog. Unable to resist interrupting the game, he leans over and says 'That's a very clever dog you have there.' 'You think so?' says the other man. 'He hasn't won a game yet.'

Are you romantic or a cynic in your attitude to animal intelligence? In the field of animal tool use it can be hard to entirely prevent temperament from intruding into interpretation. A balanced and scholarly review published in 2004, comparing the intelligence of members of the crow family with apes, had this to say of New Caledonian crows. They 'display extraordinary skills in making and using tools'. At least as far as the 'making' goes, I just can't accept that this is true. When interpreting evidence of inspiration and innovation too, I think we should also be very careful in dishing out accolades to animals. We should remember the power of genetically determined predispositions and developmental guidelines. Even some kinds of behavioural flexibility may be genetically programmed contingent responses. Let me give you an example from my own research.

The caddis larva *Lepidostoma*, which cuts the standard leaf panels to create its box girder case, also shows adaptive flexibility in determining the length of individual panels. Normally, as I have explained, the joints between panels in one side are half a panel length out of phase with those of the neighbouring sides and in phase with those on the opposite side. This gives the case additional strength. A number of years ago I carried out an experiment in which I cut back all four sides at the anterior of the case so that they were level. The response of larvae was to add panels to the front of the house of varied length on the four sides, so as to restore the out-of-phase relationship between panel joins in neighbouring sides. I would be surprised if that was other than a genetically determined response.

Psychological tests on captive chimpanzees show that, in spite of the appearance in natural, tool-using populations of an understanding of how a tool works, any reasoning process is less complex than our own. We are able to mentally represent unobservable causes and therefore devise abstract conceptual strategies to solve problems. Chimpanzee reasoning, it is argued by Daniel Povinelli in his excellent book *Folk Physics for Apes* (2000), is based upon the tangible and observable. They have, he reasons, in some respects a more accurate view of the nature of the world about them than we do. This may allow them to reason very effectively that one event leads to another without our understanding of why this is so.

Povinelli's conclusion reinforces the continuing theme of this book, that there may be effective ways of achieving behavioural goals that seem complex to us, by simpler means and with simpler brains. Here, our comparison is not between the large and the small brained but between ourselves as the large-brained party and the chimpanzee as our somewhat smaller-brained challenger.

Folk Physics for Apes is also an interesting book for its explanation of the distribution of tool use among the apes. You will remember that, apart from the chimpanzee, there is little other tool use by apes in the wild. In captivity, however, orang-utans rival chimpanzees in their use of tools, using leaves as sponges to obtain water, and sticks to reach food and even lever open cage doors. It had been coincidentally recorded that, among the apes, chimpanzees and orang-utans are the only two species that respond to seeing their image in a mirror as if they recognize themselves. Poninelli, being of the less generous school of interpretation, does not credit chimpanzees and orang-utans with a psychological concept of 'self' but rather with a recognition that everything that is true of their mirror image is true of them. More interesting, in the context of this book, is the link he makes between self-recognition of a mirror image in these two species and evolution of tool use. The link is body weight.

Gibbons, at 7 to 10kg, are much the same weight as many monkey species and so are able to swing from branch to branch with elegant but relatively stereotyped movements. As a consequence, argues Povinelli, they do not have the awareness of body posture required for effective tool use or tool making. Orang-utan females are much heavier (30 to 50kg for females, up to 80kg for males). Their movements are very different, employing the grasp of both hands and feet, and the subtle, flexible redistribution of body weight to manoeuvre skilfully from one tree to the next. So orang-utans have evolved a highly developed awareness of their body movements and posture (*kinaesthetic self-awareness*). This gives them the capacity to make and use tools. In the wild they don't show this because they are always in the trees, but in captivity they do, as they spend much more time on the ground with their hands free. Chimpanzees, according to this explanation, are also in the size range where this postural self-awareness facilitates movement through the trees. However, they also spend much time on the ground, where they can take advantage of their ability to make and use tools. But why are gorillas so rarely tool users? The reasoning here is that, having evolved into the heaviest of the apes and not being adapted to life in the trees, they do not possess the kinaesthetic self-awareness to become effective tool users.

This is an ingenious explanation, linking two seemingly separate abilities, tool use and response to one's mirror image. I do, however, have a couple of misgivings about it, one particular and one general. The particular one is that I would have imagined that the bonobo, which is about the size of a chimpanzee and closely resembles it physically, ought to be a more habitual natural tool user, but seems not to be in the wild. Some evidence from a captive bonobo, Kanzi, shows similar abilities to captive orang-utans. Kanzi, given sharp stone flakes, copied a human in cutting a string to obtain treats. Also, having seen the stone flakes being made by a human expert striking a stone in a precise way, he apparently invented his own 'flaking' technique of smashing a stone on to the hard floor.

My general, and more important, concern of the self-awareness tool-use explanation is that once again tool-related behaviour is segregated from other object manipulation. Gorillas, orang-utans and chimpanzees all make bed nests, either in trees or on the ground, in which to spend the night. This is habitual behaviour. A chimpanzee generally makes a new nest each night, and may also make a less elaborate day nest for an afternoon nap. Let's see a detailed comparison across these of their nest structures and nest-building behaviour, along with their tool-related behaviour. Also, what about the elaborate food handling manipulations of gorillas (Chapter 1, page 11), in which they fold spiny leaves to protect their mouths when chewing? Is that not evidence that they have the manipulative skills and spatial awareness to make tools?

How is the 'tools, animal intelligence' hypothesis looking? Well, it is apparent from invertebrate examples that tool use and even tool making can be shown by animals with very small brains exhibiting largely genetically determined behaviour. As to the difficulty of making and handling tools, there is no animal tool that looks particularly complicated when set against the multiplicity of other things that animals make. Evidence from the ape and bird tool makers seems to indicate some understanding of what they are doing, albeit simpler than our own. Careful experiment is beginning to tease out more precisely what that is. Nevertheless there is also evidence from bird tool makers, in particular, of strong genetic predispositions. What, in that case, can be said in support of the 'tools are not often useful' hypothesis?

A tool must be held in order to be used. We should therefore ask the question of a tool-using bird, for example, would it not be better off employing its beak directly to obtain food? We should similarly ask that question of tool-using chimpanzees or bugs. Mouths and legs are exactly the organs used in most kinds of building, as we saw in Chapter 3, but, except for tool use, this is only during the construction process. Once the house or web is built, it carries on doing its job of extending the builder's control of the environment, while the builder's legs and mouth are free to do other jobs. A tool ties up the use of the mouth or limb for the whole time that it is being used, and extends the animal's spatial influence only marginally. So how much benefit can a tool offer?

The invertebrate tool users and tool makers show us that there is a small-brained route to having these abilities. However, the number of invertebrate species showing tool behaviour is very low, particularly bearing in mind that the total number of invertebrate species runs to millions. This lack of success of invertebrate tool users is very striking when set against invertebrate builders as a whole. The evolution of nest building in solitary wasps, for example, seems to have been important in the evolution of social life in wasps and assisted their invasion of a variety of new habitats. Nest building has in fact probably had a central role in the evolution of social insects generally. In any case, there can be no doubt that the evolution of web building in spiders was a major spur to their evolutionary diversification. Also, we have yet to fully appreciate the importance of web-building spiders on the ecology and evolution of flying insects. By comparison, the influence of tool use on invertebrate evolution seems to have been virtually zero. Was that because these animals were not intelligent enough? From the evidence we have looked at in this chapter, that argument seems weak.

What of the importance of tool use in higher animals other than ourselves and its influence on their evolution? There is no single species other than ourselves that is absolutely dependent upon tools. Most tool use in birds and mammals is concerned with food gathering. So, bearing in mind that all wild populations of chimpanzees appear to use tools, what proportion of their food do they obtain by using them? Surprisingly, we don't yet have detailed information, but two very experienced field primatologists, Bill McGrew and Richard Wrangham, have given me a general estimate for two sites in Uganda. In one (Gombe), 'termite fishing' using a fine plant stem to probe into the mound, occupies about 15 per cent of waking hours during the three months or so per year when it is possible. In the other site (Kanyawara), the percentage of feeding time over the year that involves tool use is less than 1 per cent. So, at neither site are the chimpanzees heavily dependent upon their tools for food.

A study on wild populations of woodpecker finches on Santa Cruz Island in the Galapagos has found that in the more humid, wooded parts of the habitat, the finches rarely use tools. However, in the dry season in the more arid part of their range, they obtain about 50 per cent of their food from tool use. That represents a substantial effect on the habitat range of at least that species, but I wonder how generally that is the case for the other examples of tool-using birds? It would be particularly interesting to have the data for New Caledonian crows.

It may be significant that both the woodpecker finch and the New Caledonian crow, the most advanced examples of tool-using birds, have evolved tool use in island habitats. Such environments may be limited in the quantity or variety of suitable food, but allow a species to evolve relatively inefficient foraging methods, protected from too many competitors.

New Caledonian crows have relatively large brains and belong to a family of birds the brains of which are generally large compared with birds as a whole. A number of crow species have been recorded using tools, and other aspects of their behaviour (in the concealing and retrieving of food, for example, as discussed in Chapter 1, page 11) show evidence of intelligence and flexibility. Why is it then that parrots, with relatively large brains and a reputation for intelligence, provide so few examples of tool use in the wild? In a rare example, palm cockatoos (*Probisciger aterrimus*) have been observed to beat the trunk of a tree with a stout stick held in the foot in an apparent courtship display; not very impressive by the standards of woodpecker finches or New Caledonian crows. Does the combination in parrots of sharp, powerful beak and feet that can be used as hands render tool use largely unnecessary?

We also have evidence from studies on captive monkeys and apes that species, not seen to use tools in their natural habitat, will show some ability to use them in captivity. For example, four Hoolock gibbons (*Bunopithecus hoolock*), set the task of retrieving a food item using a simple, toothless rake, were all successful within about one minute, in their first trial and without any instruction. Vervet monkeys (*Cercopithecus aethiops*) and cotton top tamarins (*Saguinus oedipus*) have both been successfully trained to use such a rake to obtain food, and in modifications of the basic test show some understanding of how to use it effectively. This evidence suggests that, had there been more selective advantage for tool use in the wild, it might now be quite widespread in monkeys and apes.

For vertebrates generally, it seems that tool use has not had important evolutionary consequences, whereas building behaviour has, as is particularly evident in the nest building of birds and the burrowing of rodents. I think that the 'tools are not often useful' hypothesis deserves further investigation. If true, it would suggest that tool use has evolved repeatedly across the animal kingdom only to disappear again, a prediction that is unfortunately difficult to test. This would make the role of tools in human evolution possibly the one, spectacular exception.

It does still seem possible that the enlargement of the hominid brain over the past 3 million years was linked to the freeing of the hand from general locomotion duties so allowing more tool use, but that is still a matter of debate. There are strong rival candidates, the Machiavellian hypothesis for one. However, if tools were really important in shaping our evolution, why did humans succeed where other species failed? My guess is that social cooperation could make the difference between tools being of marginal value to being of significant benefit. Hunting as a group, even when only one or two individuals have stick weapons might make a difference to the size of prey that can be killed. However, let me return to the fundamental error, as I see it, of isolating tool behaviour from other construction behaviour. As nest building is shown by all great ape species, it probably has a history in hominid evolution at least as long as tool use. Social cooperation and handy hands could also lead to enlarged shelter building or the making of defensive fences or ditches. These could have been of significant benefit and changed the world of the hominid builders just as much as tools.

So, what sort of magic is tool use? Well, it is part illusion, but let's just enjoy that bit. However, understanding how we became large brained, intelligent and masterful is important. Tools may have had a significant role in it. We also still have plenty to understand about bird and primate tool users. I am on the lookout for new and interesting results.

8

Beautiful Bowers?

I've no hesitation in calling it a work of art, even though it is an illustration in a bird book. The very size of this illustration does, however, suggest that this is a rather special bird book; it is 54cm high and 68cm across, covering a whole double-page spread. It is Plate 8 in Volume 4 of John Gould's *Birds of Australia*, published in 1848. It shows a male and female of the spotted bowerbird (*Chlamydera maculata*) (Figure 8.1).

Both birds, illustrated lifesize, have similar mottled brown plumage except for a flare of violet-pink feathers on the back of the male's neck. But there is a lot more to the picture than that. What we are also shown, laid out in beautiful detail and—as we now know with remarkable accuracy—is the courtship bower of the male; that is, the structure he builds to assist him in attracting a female. The picture is a tribute to John Gould both as a naturalist and a publisher; however, the 'del. & lith. by J. & E. Gould' (delineation and lithography) in the bottom left of the plate is a reminder that substantial credit should also go to Elizabeth, John Gould's wife, as an artist in this great work.

The bowerbird family (the Ptilonorhynchidae) is a small one, confined to Australia and New Guinea and with no more than twenty species, in most, but not all of which, males build bowers.[1] In virtually no other species outside this family do males build specialized

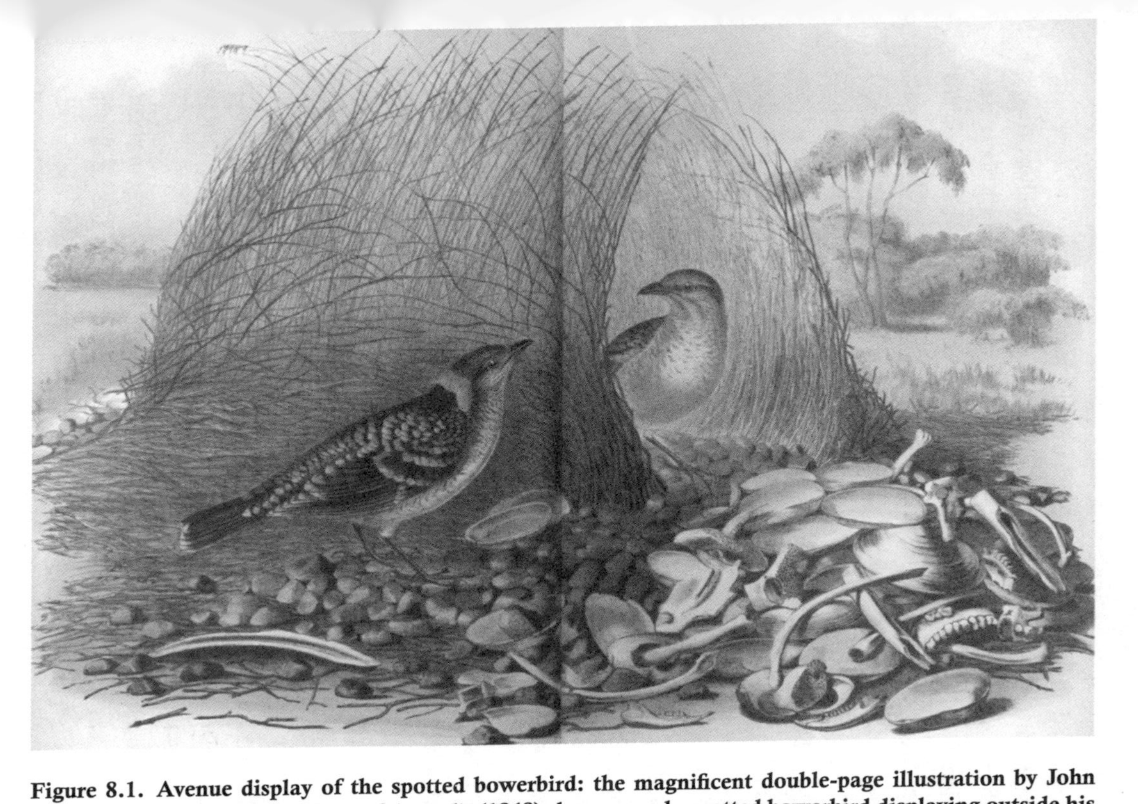

Figure 8.1. Avenue display of the spotted bowerbird: the magnificent double-page illustration by John and Elizabeth Gould in his *Birds of Australia* (1848) shows a male spotted bowerbird displaying outside his bower, while the attentive female stands inside.

Museum of Educational Heritage, Tamagawa

structures to aid their courtship and none builds a structure of comparable complexity.

A male spotted bowerbird will construct an avenue mostly out of collected fine, dry grass stems, upright but curving in towards the midline as they rise. It is in this avenue, as Gould accurately portrays, that a female stands alert and attentive as the male displays outside. Gould also depicts a cascade of smooth, knobbly black pebbles spilling out from the mouth of the avenue in front of the female, beyond which is a jumbled pile of mammal bones and bivalve (clam) shells bleached white in the sun. This does not look like, and is not, a nest. The spotted bowerbird's nest is a twiggy cup built in a tree later and by the female alone. The bower is simply a device for attracting females.

Gould, as it now appears, had to compact all this detail somewhat to fit it even on to the generous double-page spread of his book. The bones and shells together can number more than 1,000 and spread 2m from the mouth of the avenue. Gould is, however, correct in portraying the separation of pale and dark ornaments—dark stones at the mouth of the avenue, white bones and shells beyond.

More than 150 years on from Gould's observations, the variety of objects available to male spotted bowerbirds for decorating their bowers has been enhanced by human carelessness. Pieces of broken bottle glass are now a popular alternative to dark stones, as well as other shiny manufactured objects—including, it is reported, car ignition keys. What does all this mean and how did it evolve? Let me also ask, since there is almost a tradition towards the end of such a book as this to raise at least one carelessly undisciplined question, is it art? All these questions are the subject of this chapter.

In one particular, Gould's illustration is actually seriously misleading. It shows a scene that is tranquil, almost static; the reality is very different. The female, possibly initially attracted to the bower by the trail of white bones leading to it, does indeed then take up position in the grass avenue, where she remains largely passive while the male displays, departing either before or after mating with him.

The display behaviour of the male is, however, frenzied, sometimes almost threatening. He rushes at the female, jumps in the air, pecks violently at display objects, sometimes picking them up in his beak and flinging them away. Accompanying this, the male's feathers are alternately ruffled then sleeked and the bright pinkish collar crest raised and lowered. All the time he also emits a stream of mechanical and musical sounds: hisses, chatters and whistles. Some of these are recognizably mimicked sounds: songs of other local birds or a dog barking. The courtship display of the male spotted bowerbird is a wild singing and dancing routine, conducted around a manufactured stage with a special viewing platform for the female. It is a display that requires some explanation.

The *avenue* bower of the spotted bowerbird is one of two basic bower designs shown by bowerbird species as a whole. The other type is the *maypole* design, so called because at the core of it there is generally a stack of twigs built around a sapling growing on the forest floor, creating a pillar.

DNA sequencing of the different bowerbird species confirms that these two design types reflect two distinct lineages within the bowerbird family. Currently it is the avenue builders that are the best studied, in particular the spotted bowerbird, as already described, and another Australian species, the satin bowerbird (*Ptilonorhynchus violaceus*).

One of the better studied maypole builders is the so-called vogelkop bowerbird (*Amblyornis inornatus*), found in New Guinea. The maypole of this species consists of a tower of horizontally oriented twigs laid round a fine sapling up to a height of 2.5m, and smeared with a whitish, sticky material. This is a large structure, being about ten times the height of the male bowerbird, and rising out of a smooth, perfectly circular arena of dead moss, stained with the dark coloured faeces of the male himself. Outside the arena are laid, in neat and discrete piles, a variety of ornaments. One example, from the Kumawa Mountains of eastern New Guinea is recorded in detail as ornamented with a pile of small black sticks, eight green leaves laid neatly side by side, a further bundle of 32 long black sticks, 203 brown acorns, 18 brown snail shells and 243 grey snail shells. Finally, leaning like little ladders against the maypole, were three straight *Pandanus* palm leaves each nearly 1m long. The total weight of the whole display was about 3kg, twenty-four times the weight of an average male.

This is remarkable enough, but in the Wandamen Mountains, 200km away from this vogelkop population, males build a bower of distinctly different design and decoration.[2] Here, the maypole, which lacks sticky material, is enveloped in a conical-roofed hut of twigs that opens on one side on to a forecourt of fresh green moss. Decorations are plentiful, and characterized not by greys and browns, but by stronger, brighter colours: blue, orange and red berries, red leaves, black bracket fungi and shiny black and brown beetle-wing cases (elytra), each type in its own neat pile.

The bowers of male bowerbirds are, as you see, extravagant in complexity and, particularly in the maypole group, in their size. The beautiful, canary-coloured male of the golden bowerbird (*Prionodura newtoniana*) of north-east Queensland, in a variation on the maypole design, typically builds a pair of stick towers round neighbouring saplings that are naturally linked by a low horizontal perch such as a forest vine (Figure 8.2). These bowers are located in traditional sites to which material can be added over successive years to create towers over 2m high, between which the diminutive male can perch on a platform, often decorated with 'beard' lichen and creamy white flowers, advertising himself with a medley of sounds mimicking local birds and insects.

In 1871, Charles Darwin, in his book *The Descent of Man and Selection in Relation to Sex* (1871), made very particular mention of the displays of male bowerbirds. Darwin had briefly visited Australia as the Beagle headed for home at the start of 1836, but may only have been aware of the complexity of male bowerbird displays when John Gould, in a talk to members of the Zoological Society of London on 25 August 1840 showed his specimens of the bowers of the spotted

Figure 8.2. Golden bowerbird on its bower: a male golden bowerbird stands on its perch between the two substantial stick towers that characterize his bower, holding a white flower to be placed as a decoration.
Michael & Patricia Fogden/Minden Pictures/FLPA

and satin bowerbirds. It seems probable that these are the very ones on which Gould's illustrations in the *Birds of Australia* (1848) are based, since we also know that among the acquisitions of the British Museum of Natural History for 1841 were two bowerbird bowers, one from each of these species. They have subsequently disappeared, possibly when, a hundred years later in 1941, the museum was damaged by incendiary bombs. Anyway, what Darwin wrote in *The Descent of Man* was: 'the playing passages of bowerbirds are tastefully ornamented with gaily-coloured objects, and this shows that they must receive some sort of pleasure from the sight of such things'.

I will return repeatedly in this chapter to the issue of *pleasure* raised by Darwin, but the main thesis of his 1871 book was that differences between the sexes within a particular species can be explained as resulting from selection pressures for obtaining mates. He proposed two such selection processes, through male–male rivalry and through female choice. Darwin's views on this aspect of the evolutionary process, so-called *sexual selection*, as I mentioned on page 131, amounts to a second mechanism for driving the evolutionary process, one additional to natural selection. Yet, at its publication, it provoked nothing like the storm that *On the Origin of Species* had in 1859, and it was a long time before it provoked serious experimental investigation. Now the situation has radically changed; the last three decades have seen an explosion of new predictions growing out of Darwin's original thesis, and a feast of new research data. Included in this has been some very good work on the displays of bowerbirds. This is firstly because the very elaborate male displays can best be explained as the result of selection through female choice, and secondly because the separateness of bird and bower allows the latter to be experimentally manipulated with little disturbance to the male himself. These manipulations involve, for example, not only taking away some of a male's ornaments but also giving him additional ones, to test the effect of this on his mating success.

An often quoted example of a sexually selected character shaped by female choice is the peacock's tail. It is an enormous structure relative to the male bird that carries it, which is erected as a giant fan behind the male when he displays to a female. When raised and spread, the tail reveals a shimmering blue-green tail topped with a hundred or more large eyespots. Females by contrast have shorter, grey-brown tails with no eyespots. So do peahens actually choose mates on the basis of the quality of their tails? Certainly males have very variable mating success, and the males that have greater success tend to have tails with more eyespots and in better condition. But I could argue that these favoured males have been selected on quite different criteria, which happens to correlate with having an impressive tail. With the aid of a pair of scissors and some sticky tape, this can be tested experimentally.

If the ends of feathers that bear eyespots are snipped from the train of one individual and added to the train of a rival, the attentions of females tends to move in the direction of the eyespots. But we need to show more because, for the tail of peacocks to have evolved to this extent, females should have obtained some reproductive advantage from choosing males with better tails. We do, in fact, have some evidence that peahens choosing males with bigger, better trains have offspring that survive better in early life. This supports the sexual selection explanation for the peacock's tail. Females choosing males with better tails appear to have more vigorous offspring.

Sexual selection theorists have now come up with a number of possible benefits to females for choosing a mate on the basis of his display, be it bright feathers or elaborate song, but these benefits can be reduced to two basic categories: direct and indirect. A peacock with a poor quality train might have parasitic feather lice. Females gain direct benefit by being able to see this in the poor quality of a male's display and avoid infection, a direct benefit. An indirect benefit is one that she won't enjoy herself but which will ensure that her progeny are more likely to experience reproductive success. Two such indirect benefits have been proposed. These are embodied in the so-called 'runaway' hypothesis and in the 'good genes' hypothesis.

The runaway hypothesis says that, should females for any reason be attracted to some male character, blueness for example, then bluer males would get more matings. The population in the next generation would then, provided that blue plumage colour was heritable, have males on average more blue than those of the previous one. As females of this generation continue to express a preference for blueness, the bluest males (the 'sexy sons') again get most matings and, with each succeeding generation, blueness in males continues to 'runaway' towards an extreme. This hypothesis does therefore help to explain how some sexual displays, in birds, butterflies or whatever, are characterized by very exaggerated body features or behaviour. However, these can also be explained by the rival, good genes hypothesis.

The good genes hypothesis argues that males, through their extreme displays exhibit their vigour and general well-being, which in turn reflects the quality of their genetic make-up. A female that mates with such a male therefore benefits not simply by having 'sexy sons', but in having high-quality offspring, male and female, which survive well and breed successfully.

So, after that little diversion into sexual selection theory, let's get back to the bowerbirds. If we are right that male bowers and their other associated displays have evolved through female choice, then we should be able to show that females are indeed choosing, and also what features of the display influence their choice. In addition, it would also be gratifying if we could confirm that females obtain some direct or indirect benefit from the choices they make. We now have good evidence that females are choosing, but currently rather little evidence on if and how they benefit from it.

In natural populations of the spotted bowerbird (Figure 8.1) it has been shown that males vary greatly in their mating success and that more successful males tend to have a greater number of ornaments in their bowers, in particular, bones and pieces of glass. Females also prefer males with bower avenues that are neat, vertical and symmetrical. This suggests that females are choosing males for indirect benefit; that is, either for sexy sons or for high-quality offspring generally, bearing father's good genes. Nevertheless there is also some evidence in the bower design of female selection for direct benefit.

I've said that the display of the male spotted bowerbird is very vigorous; in fact it is probably the most vigorous of any bowerbird species, so one possible function of the bower avenue is to give the female some security, particularly during the initial frenzy of the male's display. This direct benefit explanation is the so-called *threat reduction hypothesis* and there are varied pieces of evidence that support it.

I should explain at this point that the orientation of the avenue in this species is not haphazard but with its axis East–West. As Australia is in the southern hemisphere, the sunniest position for the male to display is on the north side of the avenue. As the female stands in the avenue, this means that it forms a barrier between her and him but the fine grasses of its walls allow her, as Gould accurately portrays, to observe him. If one of the sides of the avenue is experimentally removed, it has been found that both the male and the female try to position themselves so as to keep the remaining wall between them. Should he in his enthusiasm find himself with no barrier between them, he reduces the intensity of his display. However, although both sexes prefer to keep the screen between them at the outset of his courtship display, results also show that females prefer males that give the most intense displays. It seems that to attract females to his display site, a male spotted bowerbird needs to give her security against possible attack from him in the early phases of his display. This then allows him to demonstrate the full intensity of which he is capable, allowing her to make a full assessment of his quality.

What about possible indirect benefits to the female, evidence that, in choosing as she does, her offspring will in some way benefit? Unfortunately we do not yet have evidence from the longevity or reproductive success of the offspring, so we need to rely on whether the males display could be an honest signal of his vigour. This is certainly supported by the preference by spotted bowerbird females for intense male displays that probably test his strength and stamina. But what about bower features such as the ornaments; can they tell the female about male quality? Does he, for example, choose rare objects as ornaments or ones that are highly perishable, such as berries or flowers? Collecting such objects would cost time and energy and so might serve as an indicator of the male's quality.

Spotted bowerbird males make use of a wide variety of ornaments, and tend also to use different ornaments in different parts of their range. A study in central Queensland found 120 different sorts of ornaments in bowers, but no evidence that rare objects were found in bowers more frequently than their occurrence in the surrounding habitat. Indeed, where at one site bleached snail shells were used as a major ornament, at another site where snails were rare, white stones were used instead.

Berries of several kinds were also found as bower ornaments but again, their use in bowers did not support the 'costly ornament' hypothesis. Berries were not selected for their rarity or for their perishability. The most preferred was found to be a green berry of a *Solanum* species (a relative of the tomato); these were both quite common and not highly perishable. But they were shown in this study area to be the best predictor of male mating success. Males with more of them secured more matings. How can this be a signal to the female of male quality? An experimental manipulation of the ornaments has turned up an interesting answer.

Suppose that, instead of taking ornaments away from the bower, the experimenter provided males with a supplement of green *Solanum* berries. You would imagine that a male would be (let's be anthropomorphic!) delighted with such a free gift. You would be wrong. Males provided with additional green berries actually removed them from the bower, restoring the number to close to their original. The explanation for this was found to be that, although a greater number of green berries was attractive to females, it also induced attacks from rival local males that damaged the bower platform and avenue. The net result was no increased mating success.[3]

It seems that in this population of spotted bowerbirds, a male displays the number of green *Solanum* berries that reflects his ability to repel rival males so, although the berries are quite common, they are an honest signal to females of a male's competitiveness, and therefore possibly a good reflection of the quality of offspring that he could sire.

Honest signal is a term in common use in the vocabulary of sexual selection because, the argument goes, in a situation where a male is trying to impress a female or indeed a rival male, the cheapest way to do it would be to cheat, by, for example, putting resources into superficial showiness to conceal poor general body condition. But natural selection should have favoured females or rival males that are sceptical and challenge the displaying male, revealing his deceit, while gullible females would leave fewer or less healthy offspring. This should lead to the evolution of signals that are truly costly and difficult to fake. The simple device of the green berries in the display of the male spotted bowerbird seems to achieve this.

So, in the study of sexual selection, male bowerbirds as well as peacocks have proved to be valuable models since both have exaggerated display features that are also amenable to experimental manipulation. There is, however, a very important difference between the two species in respect of what a female is choosing. The peahen is choosing beautiful feathers; the female spotted bowerbird is choosing a beautiful brain.

Oh, my goodness! I do apologize for such an outburst of intemperate language! I have so far avoided any suggestion that a female bird might consider a male's display to be 'beautiful'. To be able to claim that, we would need to demonstrate what Darwin was claiming in *The Descent of Man* (1871) for bowerbirds and their displays, that they 'receive some kind of pleasure from the sight of such things'. For most of the twentieth century, for a scientist to say such a thing would have been condemned as philosophical speculation or dismissed as the onset of senility, the 'pleasure' being considered not simply unlikely but also unmeasurable. The study of animal behaviour to establish its scientific credentials needed to show objectivity and investigate only the measurable. Animal pleasure was a lion's mouth into which we hesitated to put our heads. As I mentioned in Chapter 1 (page 9), Donald Griffin's (1976) book *The Question of Animal Awareness* changed all that. I am now emboldened to believe that not only might bowerbirds feel pleasure but also that we might be able to obtain objective evidence of it.

Anyway, let me try again to explain, this time in unexceptionable language, how selection by female choice is in an important way different for peacocks and male bowerbirds. Peahens are selecting the most elaborate male feathers; female bowerbirds are selecting the most elaborate behaviour (bower building, dashing about, vocal mimicry). Behaviour is organized in and generated from the brain.

Female bowerbirds are exerting selection pressure on male brains. Since, in the case of a higher vertebrate such as a bird, the brain is already a complex and sophisticated organ, the implications of this selection are very interesting. Before we end, I shall come back to it. But put aside bowerbirds for the moment and consider chimpanzees and art.

In 2005 three paintings attracted some media attention when they came under the hammer at a highly reputable art auctioneers in London. They were by the chimpanzee Congo, originally sold at an exhibition of his work in 1957 by his agent, the behavioural scientist turned popular science writer, Desmond Morris.[4] I don't know what the paintings originally sold for, but this time they fetched £12,000. I feel quite comfortable calling Congo's work 'paintings' because they are after all images made with paint, applied in this case to paper. Congo is thought to have done a few hundred in the 1950s, achieving some reputation and commercial success, but are they art?

Some of Congo's work was very favourably received when it first appeared, by at least some prominent art critics and artists. However, it was presented as 'abstract' work and, as sceptics have since pointed out, in the 1950s and 1960s American abstract expressionism, typified by, for example, the work of Jackson Pollock, was very much in fashion, providing art works alongside which Congo's could readily be compared.

A few other apes including gorillas and orang-utans, as well as other chimpanzees, have subsequently undertaken artistic careers, some having human agents with more commercial than scientific interest in their protégés. However, Desmond Morris, although keen to publicize Congo's painting abilities, did not start with this intention. Congo was initially introduced to various play objects among which there happened to be a pencil. When he took an interest in making lines with it, Desmond Morris, with his background in animal behaviour research and an enthusiasm for art, was curious to see what Congo would do with paints. Desmond Morris interpreted Congo's paintings as deliberate creative acts because, he claimed:

(1) they had elementary composition; (2) Congo chose the colours; and (3) he apparently had a view on when a painting was finished, producing a tantrum if the paper was removed 'prematurely' and ignoring encouragement to add more paint when he considered the work 'complete'.

However, I could provide a much more cautious interpretation of Congo's behaviour by saying that he threw a tantrum if his toys were removed prematurely, but did not care once he was bored with them. Interpretation of ape painting does raise problems, but we should rightly be interested in it because humans produce works of art and chimpanzees are our nearest living relatives. This leads me further into alien territory by asking what do humans regard as art and what is an artist?

I have a personal weakness for studio pottery. My, now grown-up, children gently mock me for getting a bit peculiar at the sight of even a plain brownish bowl. I am repeatedly impressed by my emotional response to such simple but, to me, sublime objects. Where did that response come from? I think that is an obviously biological question.

My question is about evolution. What are the biological roots of my response to the sight of a beautiful brown bowl? In asking it, I am implying that at least some aspect of my sensitivity is enhanced by my genetic make-up and that the genes involved in this response have been subject, possibly over very long periods of time, to natural selection, even sexual selection. This view seems to provoke two kinds of vigorous objection, one concerning human nature generally, and the other concerning specifically human art and culture. First, the nature of human nature.

You are probably aware that the Darwinian theory of evolution was embraced in the nineteenth and twentieth centuries by certain political philosophers as a justification for human selfishness and the perpetuation of social inequality. Ever since, some sociologists have attacked biologists' claims that aspects of our social organization may be subject to genetic influences as tainted by the political prejudices of Social Darwinism. This spectre was raised again in the 1970s in response to the publications of E. O. Wilson's (1975) *Sociobiology*[5] and Richard Dawkins's (1976) *The Selfish Gene*,[6] the accusation being that these books promoted an attitude of *biological determinism*, where the 'evil that men do' is regarded as an unalterable consequence of our inheritance.

I don't want to go over that ground again. Biologists have since repeatedly explained that genes interact with environmental influences in development and that, in much of human development, the scope for environmental influence is great.

The other objection, the one concerning genetic influence on art and culture, is that the human expression of these is so varied throughout the world, and so lacking in any universal criteria of beauty or quality as to be entirely culturally determined and without any genetic basis. There is also some political baggage that accompanies this, which centres on the use of the word 'art'. There seems to be a fairly widespread acceptance that the concept of 'art' as some kind of intellectually refined expression of human creativity is a fairly recent Western notion, as recent as the last 200 years perhaps. It is therefore open to the criticism that aesthetic appreciation or criteria of beauty are standards that those of a European tradition seek to impose on the world through a kind of cultural imperialism.

In spite of these views there are also anthropologists and sociologists who not only readily accept our animal origins through evolution, but also feel it worth exploring without prejudice whether that evolutionary history can enlighten what I will call our behavioural biology. One such person is Ellen Dissanayake who published in 1988 a book entitled *What is Art for*?,[7] and in 1995, *Homo Aestheticus*.[8] The title of the latter is interesting because it offers an answer to the question of who is an artist. Her answer is: *we are all artists*. We are a species, in contrast it seems to all others, universally preoccupied with artistic expression. Go into virtually any household in any part of the world and you will find ornaments. Inspect the most minimal baggage of some nomadic pastoralist or hunter-gatherer and you will find at least some utilitarian object embellished with geometric designs or figures of animals. That, of course, is just talking about visual art. What about the all-pervading presence of music-making, using a multitude of devices, not to mention the human voice?

Ellen Dissanayake asks the same question that I do when I contemplate a piece of pottery: why do these things give pleasure, even ecstasy? This leaves any debate about the definition of 'art' in modern, Western civilization to others, and concentrates on the possible biological origins of our universal artistic creativity. Dissanayake talks in the language of an evolutionary biologist in saying that 'the behaviour of art... is a biologically endowed proclivity of every human being'. Her hypothesis for the biological advantage for artistic expression is that it has the quality of 'making special'. This, she argues, can be seen in the widespread use of artistic expression in ritual concerned with ordinary experiences (birth, illness, hunting, death) to create a sense of the extraordinary, its benefit being to enhance group cohesion and a feeling of control over events. Hers is not the only evolutionary explanation of human artistic behaviour. Another hypothesis, for example, is that our remarkable creativity is a by-product of being able to imagine alternative future outcomes of events—something that could well have been of benefit to a social hunter-gatherer.

Where should we look for evidence of the biological evolution of our artistic expression and its possible selective advantage? Most obviously, to our nearest living relatives, the apes. What artistic expression do they show and does that tell us anything about ourselves?

This is where my evolutionary biologist's approach comes up against a big problem—apes, it seems, have very little to tell us about art. There is virtually no evidence of any apes exhibiting in the wild, behaviour that might be loosely interpreted as artistic expression. This leaves us with a clutch of captive, ape-painters of uncertain significance. I say virtually no evidence from the wild but, I should at least give a mention to 'Ako's knot'. Ako was a young adult female chimpanzee that was observed, a day after her group had caught and killed a red colobus monkey, wearing a long strand of colobus skin round her neck. Later the same day she discarded it, allowing it to be collected and examined. This revealed that the ends had been tied together in a single overhand knot to create the equivalent of a necklace.[9] What significance can be attached to this object? Unfortunately, very little. Its very uniqueness attests to the fact that it does not contribute to or reinforce any similar pattern seen in wild chimpanzees. We have no idea whether the knot came about through deliberate intent or accidentally, or even if Ako herself made it.

I want to digress into literature for a moment in order to draw attention to the way in which we interpret art. 'Reflections are images of tarnished aspirations.' How do you feel about this quotation? It was written by a computer.[10] One with a language program equipped with a vocabulary, some basic rules of grammar and some additional instructions, e.g. how to conjugate verbs and form plurals. The machine meant absolutely nothing in what it wrote, but to us the sentence has a bitter-sweet evocation of regret and loss.

How about something more surreal. 'In a half bright sky an insect wraps and winds a chain, a thread, a cable around a sphere of water'? Same computer program, but this time sounding remarkably like a pastiche of Walt Whitman: 'A noiseless patient spider, I mark'd where on a little promontory it stood isolated. It launch'd forth filament, filament, filament, out of itself. Ever unreeling them, ever tirelessly speeding them.' The point I am making is that our minds are designed to extrapolate, to fill in the gaps, to look for meaning and even find it where none was intended. In the realm of visual images we do the same. A few black lines on a flat piece of paper are sufficient for us to see a landscape of hills and cottages.

So where does that leave us? We are an extraordinary species. One feature that makes us so is our preoccupation with artistic expression. I have argued that this raises legitimate questions for evolutionary biologists on the origins of that singular characteristic, but when we look to our nearest living relatives for insights we find essentially nothing. That seems to leave us facing a dead end in our enquiry, except I suggest that there is another animal model for us to study, the bowerbirds.

It is natural to think of our nearest living relatives as the appropriate model for coming to understand our own biology. That is frequently the case; however, it could be any species that shares the particular aspect of biology that we are interested in. Much of our understanding of genetics, after all, comes from breeding experiments on the fruit fly *Drosophila*. It has chromosomes, and on those chromosomes, encoded in the DNA, are genes; exactly the same system of inheritance as ourselves. So I suggest that the study of bowerbird courtship displays might tell us something about the evolution of human aesthetic senses because both bowerbirds and humans create complicated display objects, and because no other type of bird or mammal builds anything comparable.

In the previous chapter I was being cautious in acknowledging that tool making was as significant a landmark in human evolution or as significant a hallmark of intelligence as popularly believed. Now I seem to be losing my critical faculties in suggesting that bowerbirds might be artists. I don't think that I am for two reasons. The first is that I am simply seeking to justify an exploration, not asserting the outcome of such an exploration. The second is that I am not looking to demonstrate that bowerbirds are or are not artists, but rather to look for evidence that will help us understand how *we* became artists.

Let me take another short detour to consider that familiar object of evolutionary debate, the human eye. It is a remarkable optical instrument. It records, among other things, brightness, colour, shape, movement and distance, but we know very well that, in the evolutionary history of that organ, these abilities were not all acquired together. To get to its current state, the capabilities of eyes were added to and refined through a long history of natural selection. By analogy, I don't suppose that the biological contribution to human artistic creativity and aesthetic judgement came as one package at one time. It will have been gradual and incremental and, most importantly, I expect that some elements now incorporated into our aesthetic judgement have an evolutionary history that long precedes it, in which they had a somewhat different biological function. Some elements of that history may be discernible in the displays of bowerbirds.

I need to explain in a rather more detailed, scientific way why I think bowerbirds are a suitable model for what I propose. It is a bit more than just saying that bowerbird males make structures that look to us like works of art. It is the problem of explaining the degree of complexity in the display. The architecture of the bower, its varied ornaments, the male's movements and his complex vocalizations—what is all that doing? It does on the face of it seem to be an extraordinarily elaborate way of saying, 'I'm better than he is.'

The problem of explaining the complexity of animal displays is not a new one, and various hypotheses have been proposed to account for it. One is simply that the message is full of redundancy. If you say the same thing in lots of different ways then no one can misunderstand your meaning. A related explanation argues that the target audience contains different categories of individuals, so that part of the display is best understood by one section of the audience, another part by another section, but the message is essentially the same to both. A third explanation is that the overall display contains several messages that might be either discrete or sequential, allowing the observer to make a separate decision on each part before the display progresses. Whatever the explanation for the complexity, a female will ultimately have to make a judgement on who is the most acceptable mate. To do this, she must have some kind of brain mechanism which assembles the information gained from observing a number of males and comparing them.

All these explanations of the complexity of male displays assume that they are seeking to convey their quality and that females endeavour to assess it. Against this I want to put up a rival which, shunning equivocation, I am calling '*the art school hypothesis*'. This says that the complexity of the male display is designed to create something beautiful. Females therefore are judging males on the beauty of their presentation. The distinction may not be immediately obvious but the art school hypothesis, you will notice, has an additional layer of explanation: the complex elements of the display together contribute to a performance that is assessed by females in terms of its beauty.

So what evidence should we seek in support of the art school hypothesis? Well, it is an assumption of the hypothesis that for a male to combine all the display elements effectively he will need to learn to become an artist, and, for a female to judge effectively the subtleties of male performance, she will need to learn to become an art critic. This allows us to make certain predictions.

Males should:

1. Take a long time to become capable of attracting females.
2. During that time, show evidence of practice and improvement.
3. During that time, also try to discover what successful males do.
4. In making bowers, possibly show artistic mannerisms.
5. In different regions or parts of their range, exhibit different culturally based preferences in the nature of the display.

Females should:

1. Show evidence of comparing male performances.
2. Show improvement with age in their ability to judge bowers.

Additionally there are two important predictions concerning both sexes.

1. Bowerbirds, male and female, should show evidence of specialized brain mechanisms not shown by other birds.
2. Males should, in the making of a bower, and females in observing a displaying male or even a bower alone, experience pleasure.

I guarantee that this last word will have caused alarm and even hostility among some. I ask you to keep your seat-belt fastened until I have provided a little more explanation.

Let's start with a reminder of the principle of *Ockham's razor* or *the law of parsimony*, which came up in Chapter 5 (page 125) in determining the most likely family tree for the evolution of swallows and martins. It is the principle that we should proceed by always adopting the simplest possible explanation that will account for current evidence. So what, in the context of bowerbird displays, would be an example of that principle?

Well, suppose we want to explain how female bowerbirds assess the quality of a male's display. I could argue that a female rates the various elements of the display to create a display score, and compares the totals of the various males to determine the winner. Alternatively, I could say that the female gathers information about all the elements of the display, feeds them into a 'pleasure' mechanism from which she obtains a sensation equivalent to the value of the display, which allows her to compare males and decide a winner. It seems that, in proposing the art school hypothesis where beauty is said to be assessed as pleasure, I have deliberately defied the principle of Ockham's razor. I am going to cite a book, *The Mating Mind*,[11] written in 2000 by the psychologist Geoffrey Miller, as evidence that I have not.

This book is about the evolution of the human brain and some of the remarkable behaviours we show that are associated with it. Miller's thesis is that sexual selection has been a major driving force in the rapid increase in brain size during hominid evolution, an explanation I touched on in the last chapter (page 183). This means males choosing females and females choosing males on the basis of the versatility or creativity of their behaviour. This could be any kind of behaviour; language is a good example. Why do we like rhyme, why do tabloid newspapers delight in appalling puns for headlines, and do the Inuit really need all those words for snow? Perhaps the joy of words came about through sexual selection.

For the visual arts and music, the same explanation could apply; these are displays of creativity generated by creative brains, which affect mate choice by both females and males. The possible selective advantage to the chooser of a mate is that a creative display by a suitor may indicate more general adaptive traits—flexibility in social relationships or foraging, for example. So how does Miller propose that these displays are measured? Well, by pleasure—because, he argues, that is a highly adaptive method of making assessments and learning from experience. In the context of sexual displays, it allows very varied aspects of a potential partner's display to be combined, remembered and compared with the complex performances of rival candidates.

Miller's support for a pleasure system actually goes beyond the context simply of choosing a mate. It could be used in assessing what is good or bad food, for example. The pleasure of the experience reinforces it and aids the learning process. Miller argues for the adaptive advantage of a *unified pleasure system*. This he also points out could, through its flexibility, aid rapid evolution in behaviour. For example, if food and song both give pleasure, then feeding might become added to a courtship ritual alongside courtship song because both features could readily be combined through the unified pleasure system. Miller's hypothesis presents a pleasure system as a simplifying rather than a complicating mechanism, overturning objections that it defies the principle of Ockham's razor. We are now ready to return to bowerbirds and evaluate the art school hypothesis.

The satin bowerbird is an avenue builder with a population distribution predominantly down the eastern side of Australia. Its avenue is made of sticks and is oriented with its axis North–South. Ornaments are placed in front of the north end of the avenue, the end where the male performs his display, while the female watches, as with the spotted bowerbird, from within the avenue.

The ornaments used by male satin bowerbirds are varied: yellow and blue flowers, snail shells and the shiny cast skins of cicadas. A variety of man-made objects also inevitably turn up, but especially blue ones, having the advantage of durability over flowers. However, the most favoured of the natural ornaments are blue parrot feathers. One study establishing blue tail feathers of the crimson rosella *Platycercus elegans* as most favoured local ornament, found that second-placed, blue bottletops shared with the parrot feathers the highest level of ultraviolet (UV) reflection compared with less favoured ornaments.[12]

An additional feature of the bower is that the inside walls of the avenue are frequently painted. The paint is generally a mixture of pulped fruit and saliva, sometimes mixed with wood ash. This is often applied with the aid of a wad of bark fibres held in the beak.[13] This, it has been suggested, is not so much a paint brush as a sort of sponge that holds the male's beak open while he dabs and wipes the dribbling paint over the two inside walls of the avenue.

A male attracts a female to his bower with a special advertisement call and, on her arrival, begins to court her, tentatively at first, until she is positioned in the avenue. Beginning his full display, the male picks up a snail shell or blue parrot feather and begins to run around, flicking his wings, puffing his feathers and bowing his head, causing the light to shimmer off his iridescent plumage that reflects strongly in the violet-ultraviolet part of the spectrum (300–420nm), all of which is visible to birds. While doing this, the male also emits loud mechanical whirring or buzzing noises. Like the displays of the spotted and vogelkop bowerbirds, this is a multimedia performance: auditory and visual, combining the actions of the male and the structure of the bower.

A chemical signal may also be involved, emanating from the paint. Females, when inside the avenue, frequently peck at its walls. That she is tasting the paint is suggested by the frequency with which males paint the walls, which is increased when there are drying winds. It has also been found that if a male has his bower walls painted for him by an experimenter, then he will increase the rate of his advertisement calls. Whatever sort of signal the paint is, it does seem to influence mating success. Males that do more painting also have a higher mating success.

Male mating success in any particular area is found to be very unequal, with a successful male securing up to thirty matings in one season; a poor male, none. However, it is not just painting frequency that affects a male's success; so also does the number of ornaments, in particular, the number of blue parrot feathers and snail shells. In addition, his success is also influenced by the symmetry and rigidity of the avenue walls.

So how do these characters indicate to the female the quality of the male? Well, we do know that females take the opportunity to compare males. Male satin bowerbirds make their bowers sufficiently close together that a female can readily move from one to another. Video cameras trained on the avenues of all local bowers, and females identifiable by combinations of coloured leg rings, confirm that any female will visit several bowers.

However, the very proximity of the bowers also means that males can visit the bowers of their rivals, bent on lowering the quality of them to the benefit of their own. A male that finds a neighbour's bower unattended will assault the avenue, uprooting its twigs, and steal valuable ornaments such as blue parrot feathers. To a female, therefore, the sight of a bower with a firm symmetrical avenue, embellished by a forecourt replete with blue feathers can be regarded as good evidence of a male's competitive ability—his ability to protect his own bower while perhaps damaging those of others.

None of this seems to provide support for the art school hypothesis. It suggests that the quality of the male display is not judged by its subtlety and refinement, so much as by its vigour and endurance. As to what the paint, wet or dry, might indicate about the quality of the male—who knows? However, it too might just be an indication of endurance. Males with more time to spare for painting are stronger. What can make that cheat proof? Well, to always show fresh paint.

What of the complexity of the display? The art school hypothesis argues that it would be difficult for a female to remember accurately all the different aspects of a particular male's display (his blue ornaments and vocal repertoire were worth respectively a '6' and a '7', but his painting was only a '4'), and then compare them at some later date with the scores other males got for their various display elements. There is then the additional complication of weighing the relative importance to them of the different elements (which is more serious, poor vocal display or poor painting?) How does she decide? It is simpler, the art school hypothesis claims, to compare the recollection of the pleasure each male engendered.

There is evidence that females find the choice of a mate a complex process in the amount of information they seek, and the time and effort they invest in obtaining it before making up their minds. Even before nest building, a female will visit a number of bowers and inspect them while males are not in attendance. She will then return to some of these bowers when the males are present and observe their courtship displays, still without mating. She then spends about a week making a nest in preparation for egg laying before returning to some, but not all, of the bowers previously visited, to be courted again by these selected males before finally mating with one of them.

The art school hypothesis also gets some support from evidence that a female's decision to return to a particular bower after nest building is affected by her age and experience. In an experiment where some males were provided with additional blue feathers in the pre-nest building, courtship phase, first- and second-year females returned preferentially to males that had been given additional feathers by an experimenter. However, females of three years or older did not. It seems that younger females rely more on the blue decorations, while older females rely more directly on the male's display.

The explanation for this difference may not, however, be evidence of a female's growing artistic interpretation with age, but simply that she is less afraid of males. It appears that, as in the spotted bowerbird, at least one function of the bower avenue is to offer security to the female while the male displays his vigorous, sometimes alarming routine. Overall mating success indicates that females prefer males that give the most intense displays; however, they are frequently startled by the intensity of the male's outbursts, possibly fearing attack or rape. As a female gets more used to the presence of the male, she indicates this by lowering her head and raising her tail in a posture resembling the mating position. This suggests that a male could infer whether or not a female is likely to make her escape from the avenue by the degree of crouching she shows. Cue an experiment with a robot female satin bowerbird!

A model female satin bowerbird was constructed that could mimic the startle or relaxed postures of the normal female.[14] When placed in the bower avenue, this elegant robot induced apparently normal courtship. By altering the crouching posture of the female, it was found that a male diminishes the intensity of his display if the robot rises to a more upright posture. However, the likelihood that a real female will show the startle behaviour diminishes with the successive courtships she experiences, allowing a male to show the full intensity of the display of which he is capable. It seems possible that the greater reliance of younger females on the number of bower decorations rather than intensity of male display may be related to their greater fearfulness in the presence of males. If this is the explanation then it supports the idea of signal duplication, one part more suited to more experienced, and the other to less experienced females.

The art school hypothesis predicts that the development of full display ability in males will take a long time. Evidence does support this prediction. It generally takes five or more years before a male satin bowerbird acquires full adult plumage and is able to attract a female. During this time, juvenile males engage in an extended process of learning. Groups of individuals of about two years old have been observed working together to make a basic platform, and single immature males will repeatedly construct and dismantle recognizable but incomplete avenues. Juveniles also seem to seek knowledge from the activities of mature males. They visit displaying males, watching their performances and, if the owner of a bower is temporarily away, will enter the avenue and inspect it.

Of course, it could be argued that the delay in reaching maturity is not to do with the time taken to learn how to attract females, but simply due to some delay in the physiology of maturation. However, in an experiment using testosterone implants, males were induced to moult into adult plumage prematurely. Their bower building and courtship displays were, however, found to be no better than untreated males of the same age, and they were tolerated less by naturally mature adult males. It seems that male satin bowerbirds are actually acquiring something through experience that takes a long time, and which therefore could be complex and subtle. I find this particularly striking if you compare bower building with nest building. I have already said that I think that more learning is involved in nest building than is acknowledged. Nevertheless such evidence as we do have indicates that birds generally can build quite effective nests at their first attempt with no previous direct practice. What on earth are male satin bowerbirds learning that appears to take several years?

It is also worth noting that, although we have much less complete information for other bowerbird species on the development of male displays, what we do have suggests that they also have a protracted learning period for males before they display effectively. Males of the tooth-billed catbird, or stagemaker (*Ailuroedus/Scenopoeetes dentirostris*), and of two of the maypole building group of species, the golden bowerbird (*Prionodura newtoniana*) and Macgregor's bowerbird (*Amblyornis macgregoniae*), are also several years old before they begin effective displays.

The occurrence of regional differences is another prediction of the art school hypothesis and, as we have already seen, such differences do occur in both the avenue building spotted bowerbird and in the maypole building vogelkop. However, the art school hypothesis requires that these regional differences result from local differences in what you might call 'taste', which is culturally transmitted, i.e., learned by each successive generation of males and females from the example of already displaying adult males. However, there are two possible alternative explanations for regional differences: that they are genetically determined (with function unknown) or that they are ecologically determined (for example, an ornament is not used in one area simply because it is unavailable).

These three rival explanations were investigated experimentally on two populations of the vogelkop bowerbird with distinctive regional bower styles of a kind described earlier in this chapter: those in the Arfak area of New Guinea, where the maypole of the bower is enveloped in a hut-like canopy and the ornaments are characterized by piles of colourful fruits, and those in the Fakfak mountains, where there is no canopy around the maypole and the ornaments, snail shells, nuts and dark fungi, are drab in colour. An experiment involved the placing of small tiles of a mixture of colours near to the bowers of males in both these regions. These proved to be very acceptable but the choice of colours in the two locations was very different. In the former of the two regions, virtually all the males made use of the red and blue tiles offered, the other colours being less preferred. In the latter region, it was black and brown tiles that proved the most popular. The conclusion therefore is that the colours of ornaments collected by males is not a reflection of availability; the males are exhibiting a preference, either genetic or cultural, although at present we do not know which.

Measurement of genetic similarity (genetic distance) between the vogelkop populations of the two regions shows that they are very similar, but, since they are not identical, this does not tell us whether or not genetic difference is responsible for regional bower styles. This obstacle could be overcome by cross-fostering eggs between populations: would fostered males bring a genetically determined style with them, or would they learn the style that was practised locally? However, the implications of such an experiment for the conservation of natural populations are too serious to justify it.

I mentioned 'artistic mannerisms' among the art school hypothesis predictions, probably conjuring up anthropomorphic images of standing back from one's masterpiece, head to one side, chin in hand, before adding a single small dab of additional colour to the canvas. A scientist looking for a bowerbird equivalent would seem to be inviting derision. However, we do have anecdotal observations of male vogelkops apparently fastidiously arranging and rearranging their piles of ornaments, and of individual idiosyncrasies or very local preferences in ornaments. We also have experimental evidence that male great bowerbirds *Chlamydera nuchalis* choose ornamants that enhance contrast with their own plumage, other bower ornaments and even the nearby vegetation. Ornaments are therefore chosen according to context, even to the extent of creating a more muted background colour against which to set off more brightly coloured objects.[15] It seems worthwhile compiling observations of this sort, both anecdotal as well as experimental, in a systematic way. Anthropomorphism can be a useful starting point for an idea. More real data might provide real insight.

This brings me to a particularly important prediction of the art school hypothesis, that bowerbirds in comparison with other similar birds should have specialized brains. This was, you will remember, based on the argument that whereas the peacock is being selected for the quality of his tail, the bowerbird male is being selected for the quality of his behaviour and hence his brain, and females for their ability to discern the relative merits of these elaborate male performances. Such evidence as we have at the moment supports this.

Measurements of brain-case capacity have been obtained from X-rays of the skulls of museum specimens. This found a positive correlation within the bowerbird family between brain size and bower complexity. This relationship was found in females as well, although it was less marked than in males. More recently studies on the actual brain anatomy of just males in five bowerbird species, failed to show the correlation between the size of the whole brain and bower complexity. However, the brain of the non-bower building species was smaller than that of the bower builders, suggesting that bower-building behaviour does necessitate a larger brain.[16] The one area of the brain where bower complexity was found to be correlated with volume of the brain structure was the cerebellum, a part of the brain known to be associated in rats with learning from observation and experience, and exploring of novel situations. These are the first steps in understanding the special features of bowerbird brains.

And so to the last prediction of the art school hypothesis, the one originally proposed by Charles Darwin, which is that males, in making their bowers, and females in viewing them, experience pleasure. Field experiments have already led to important discoveries on the basis of female choice. To gain more insight into what bowerbirds are thinking and feeling, the most promising approach may now be that of experimental psychology. In fact, researchers on animal welfare are already investigating whether farm animals can be raised in conditions that are not simply free from stress, but also provide pleasurable experiences. An example of this approach is one based on *consumer demand theory*, a concept that has come from the study of human shopping habits. The purchase of an item can be described as 'elastic' or 'inelastic', depending upon the degree to which consumers will continue to buy it when its price goes up. If the price of bread doubles, for example, you still might continue to purchase much the same amount because you regard it as essential—an inelastic response. However, if the price of scented candles doubled you might well give up buying them altogether, an elastic response.

This principle was used to give a measure of how much pigs like to have social contact, not actually much social contact in the experiment I am about to describe—a few seconds snout contact in fact. Pigs learned to press a plate with their snout to obtain this modest social experience. Once a pig had learned to do this, it discovered that it had to press the plate more times just to get the same few seconds of social contact—the price had gone up. It was found that as the price went up, so the 'purchases' of social contact gradually went down. The response was elastic, but not nearly so elastic as plate pressing for a chance to look into an empty room. Food purchase turned out, as expected, to be quite inelastic. By comparing the elasticity of pigs' responses to these three commodities (company, space and food), it was seen that pigs do like to have social contact and are prepared to pay a moderately high price even for very brief social contact. This does not tell us anything directly about pleasure, but it is a flexible way of investigating an animal's preferences. Using this approach, a female bowerbird could, for example, be asked to express her preference for subtly different combinations of bower ornaments.

Some animal welfare research on birds does, however, provide more suggestive evidence of pleasure. Domestic hens will learn to perform an operation (push open a door) to get to a nest box to lay an egg. If delayed in their efforts, they become restless, show signs of agitation and will work very hard to get to this favoured egg-laying site. This is interpreted as trying to reach a goal to relieve negative feelings. However, in getting to a sandy patch where they can dust bath, hens behave differently. They will work quite hard at pushing open doors to get to it, whether they have been recently deprived of the opportunity or not. Then, having reached the sand, they may not even bother to dust bath. This cannot be interpreted as trying to relieve some negative feeling, but has tentatively been interpreted as evidence that the hens are anticipating the pleasure of a dust bath when they do feel like having one.[17] This may not directly translate to a test on pleasure in bowerbirds, but does illustrate the increasing power of the experimental psychology approach to help us understand animal feelings.

A completely different approach to understanding what bowerbirds are thinking and feeling would be to monitor brain activity directly, then relate that to what we know of the function of different parts of the bird brain. This would of course be a lot more intrusive, therefore less easy to justify. However, there are two body scanning techniques, currently used on humans, which allow the activity of the brain to be monitored in some detail without the need for the insertion into it of wire recording electrodes. Neither of these is currently adapted for studying the brain of a bird, but this does seem a realistic prospect. They are *Positron Emission Tomography* (PET) and *functional Magnetic Resonance Imaging* (fMRI). These need a little explanation.

Different parts of our brain have different functions so, when we are engaged in a certain activity or thinking about particular things, only some parts of our brains are involved. These two imaging techniques take advantage of the fact that blood supply to the brain is dynamic, responding to local demand for oxygen and nutrients.

When one location in the brain becomes activated, there is a surge of blood flow, when it is inactivated again, blood flow slows. These scanning techniques can display where and when these blood-flow surges are occurring.

PET depends upon the injection or inhalation into the human subject of a radioactive substance such as water or carbon dioxide where the radioisotope used (e.g. ^{15}O or ^{11}C) has a half-life of only a few minutes. The radioactive marker then becomes concentrated in a metabolically active area of tissue. This, in the brain, will be where the blood flow has become enhanced. Less invasive, because no radioactive substance needs to be administered, is fMRI. It detects local oxygen demand from the difference in magnetic resonance between oxygenated and deoxygenated blood. It was this technique that was used to show that a car accident victim was still able to understand and think in spite of being unable to move or communicate in any way. When asked to 'imagine herself playing tennis', the areas of the brain that 'lit up' in the fMRI scan were the same as for normal human subjects.

The basic MRI scan technique, which does not reveal any dynamic processes such as changes in blood flow, has been used to study details of the brain anatomy of anaesthetized birds as small as a canary (weight 20g, compared with 128g for the spotted bowerbird). However, there is a major constraint to the study of brain function in relation to behaviour: the head of the subject needs to be kept still during a scan. Would a bowerbird tolerate this, with or without sedation? There is clearly an ethical issue here but, if the bird is stressed in this situation, this would affect the result, making the experiment not only unjustifiable but also meaningless. However, if a bowerbird could gradually become familiar with and tolerant of the procedure, then its brain activity could be scanned while it was shown specially edited images of male displays. This would reveal which areas of the brain were associated with observing male display. As far as the art school hypothesis is concerned what we would hope to see is activity in parts of the brain that can eventually be related to pleasure. This approach is already being applied to the human experience of pleasure.

The point of this chapter has been to explore whether bowerbirds could be a model for gaining some understanding of how we came to have a sense of beauty. It is time to return to thinking about us. Humans may obtain pleasure in seeing a person, a landscape or a decorative object created by another human. These people, places and things are beautiful. According to Geoffrey Miller, this pleasurable sensation is generated by a universal pleasure centre when it receives a complex of visual information. Since humans can report on what gives them pleasure, I can ask you directly what you judge to be beautiful and why. If the human judgement of beauty is in any degree shaped by our genetic make-up, then the pleasure you experience from looking at a certain human artefact, for example, should be shared by a wide range of people across any culture. But can you suggest any criterion for visual beauty that might be so universally applicable? I want to consider one possible candidate: it is symmetry.

Humans of both sexes have been repeatedly shown, in psychological tests, using subtly modified images of the human face, to prefer more symmetrical features, even when unaware of it. Symmetry is of course a well-known theme in art and architecture, bilateral symmetry being a particularly common one. The architectural masterpiece the *Villa Barbaro* (1558), designed by Andrea Palladio, is a formal celebration of symmetry. At its centre is a frontage like a Roman temple, the apex of its triangular pediment defining the axis of symmetry. Matching, low arcaded wings are then bracketed between restrained, identical pavilions.

You may object that some visual art forms deliberately celebrate asymmetry, in those highly asymmetric Chinese landscapes of the late twelfth century Ma-Xia school for example, where the rugged hills in the bottom left fade to empty space in the top right. This is a justified criticism, but I don't think a damaging one for the argument of universal criteria in aesthetic judgement. I don't want to imply a judgement as simple as 'bilateral symmetry is beautiful and asymmetry is not', only that bilateral symmetry is one significant reference point that interacts with others in the determination of the pleasure of a visual experience.

Coming from a biologist, these observations on art and architecture may seem like presumption or more likely just ignorance, but biologists have a particular interest in symmetry. We know it as a sexually selected character with a very ancient history. We know, for example, that the long, forked tail of the male barn swallow influences mate choice. Females prefer males with longer tail streamers, but they also prefer males with streamers of equal rather than of unequal length; they discriminate against asymmetry. This discrimination against body asymmetry seems to be widespread, also occurring in swordtail fish (*Xiphophorus cortezi*) for example.

Departure from the bilateral symmetry that characterizes the great majority of animals is an indicator of faulty development. The basis of such faults could be genetic or a poor environment during growth. In either case, a female, in looking for the most symmetrical partner, may be ensuring the best possible quality in her offspring. Please note, I am not saying that a female swordtail fish regards a symmetrical male as beautiful, or indeed that it is a pleasure for her to encounter one. Simply that she makes a choice based in part on the male's closeness to bilateral symmetry.

Consideration of sexual selection and symmetry brings us back to the bowerbirds because females of the avenue-building species, the spotted bowerbird and the satin bowerbird, both prefer the walls of the bower avenue to be symmetrical. This probably evolved in the ancestry of bowerbirds, when the criterion of symmetry that was initially applied by females just to the bodies of males later came to include the bower. If my reasoning seems unconvincing, I should add that we now know that, in a variety of animals, where a particular colour or shape had evolved as a way of identifying food, it has later evolved to be part of a male's display; the male's display has evolved to exploit a preference already present in females. This phenomenon is called *sensory bias*. The extension of the criterion of symmetry in female bowerbird mate choice from a male's body to his bower does, in that context, seem very possible. If, as I suspect, the human preference for symmetrical faces has an ancient and similar biological origin, it seems worth looking for evidence that humans came to use this as a criterion in judging the beauty of the objects they made as well, particularly if those objects were ornaments that influenced mate choice.

Clear evidence of artistic endeavour in humans is only very recent. The famous cave paintings from Lascaux in France date from only about 25,000 yeas ago. Excavation at a cave site in South Africa near Cape Town unearthed forty-one small snail shells, each with a hole pierced in it, as if they were a collection of beads. They have been dated at 75,000 years old. Lumps of ochre clay that were possibly used as body pigments have been found in African cave deposits which date 120,000 years ago. Is this the earliest evidence of human decorative art work?

The earliest known stone tools are, as we saw in the last chapter, a lot older. Simple stone core tools associated with *Australopithecus* species date from 2.5 million years ago. Regularly shaped, stone hand-axes, associated with species of *Homo*, date from as long ago as 1.4 million years, possibly a little earlier, but stone tools like this continued to be made until as recently as 75,000 years ago, more than a million years of hand-axes. By 800,000 years ago *Homo heidelbergensis* in Europe was making carefully worked, symmetrically almond-shaped hand-axes that, to us, are a pleasure to behold. Was that intended by their makers? Marek Kohn, in his (1999) book *As We Know It*,[18] suggests that these hand-axes which, when they first appeared were intended by their makers as utilitarian objects, eventually became not actually hand-axes at all but display objects. The reason was that they were subject to sexual selection.

Kohn's speculation is that it was male hunters that initially made the hand-axes, allowing females to judge their skill, but later the display of skill became an end in itself. The evidence for this is that many of these later hand-axes show excessive detail in the workmanship, have cutting edges quite undamaged by use, and are generally markedly bilaterally symmetrical. A conspicuous example is the so-called 'Furze Platt Giant' from the Thames Valley gravel beds.[19] It is over 1ft (321mm) long, and with a carefully worked cutting edge all round its perimeter. As a hand-axe this does not look used or useful; as a display object, it looks to us at least, magnificent.

Can the study of bowerbirds help us understand how we came to recognize and enjoy beauty? Well, I'm convinced that some elements of that capacity will be found in non-human animals, and bowerbirds are a good place to start looking for them. We should add this to the other reasons given through this book for studying structures built by animals.

Notes and References

GENERAL BACKGROUND READING

Animal architecture is not generally reviewed and discussed as one subject. As that is what I have always been interested in doing, my three monographs do provide further more detailed background to all the chapters in this book. They are:

Hansell, M. H. (1984). *Animal Architecture and Building Behaviour*. London: Longman.

Hansell, M. (2000). *Bird Nests and Construction Behaviour*. Cambridge: Cambridge University Press.

Hansell, M. (2005). *Animal Architecture*. Oxford: Oxford University Press.

Similar in coverage, but richly illustrated and intended for a popular readership is:

Von Frisch, Karl (1975). *Animal Architecture*. London: Hutchinson.

REFERENCES

Chapter 1

1. Griffin, D. R. (1976). *The Question of Animal Awareness*. New York: The Rockefeller University Press.
2. Byrne, R. W., Corp, N., and Byrne, J. M. E. (2001). Estimating the complexity of animal behaviour; how mountain gorillas eat thistles. *Behaviour* 138: 525–57.
3. Jackson, T. P. (2000). Adaptations to living in an open arid environment: lessons from the burrow structure of the two southern African whistling rats, *Parotomys brantsii* and *P. littledalei*. *Journal of Arid Environments* 46: 345–55.
4. Hölldobler, B. and Wilson, E. O. (1990). *The Ants*. Berlin: Springer-Verlag.

Chapter 2

1. Löffler, E. and Margules, C. (1980). Wombats detected from space. *Remote Sensing of Environment* 9: 47–56.

2. Neal, E. G. (1986). *The Natural History of Badgers.* London: Croom Helm.
3. Groenewald, G. H., Welman, J., and MacEachern, J. A. (2001). Vertebrate burrow complexes from the early Triassic Cynognathus zone (Driekoppen Formation, Beaufort Group) of the Karoo Basin, South Africa. *Palaios* 16: 148–60.
4. Roberts, R., Walsh, G., Murray, A., Olley, J., Jones, R., Morwood, M., Tuniz, C., Lawson, E., Macphall, M., Bowrery, D., and Naumann, I. (1997). Luminescence dating of rock art and past environments using mud wasp nests in Northern Australia. *Nature* 387: 696–9.
5. Zeibis, W., Foster, S., Huettel, M., and Jorgensen, B. B. (1996). Complex burrows of the mud shrimp *Callianassa truncata* and their geochemical impact on the seabed. *Nature* 382: 619–22.
6. Büttner, H. (1996). Rubble mounds of sand tilefish *Malacanthus plumieri* (Bloch, 1787) and associated fishes in Colombia. *Bulletin of Marine Science* 58: 248–60.
7. Martinsen, G. D., Floate, K. D., Waltz, A. M., Wimp, G. M., and Whitham, T. G. (2000). Positive interactions between leafrollers and other arthropods enhance biodiversity on hybrid cottonwoods. *Oecologia* 123: 82–9.
8. Bordy, E. M., Bunby, A. J., Catuneanu, O., and Eriksson, P. G. (2004). Advanced Early Jurassic Termite (Insecta: Isoptera) nests: Evidence from the Clarens formation in the Tuli Basin, southern Africa. *Palaios* 19: 68–78.
9. Dejean, A. and Durand, J. L. (1996). Ants inhabiting *Cubitermes* termitaries in African rain forests. *Biotropica* 28: 701–13.
10. Odling-Smee, F. J., Laland, K. N., and Feldman, M. W. (2003). *Niche Construction: The Neglected Process in Evolution.* Princeton: Princeton University Press.

Chapter 3

1. Turner, A. (1989). *A Handbook to the Swallows and Martins of the World.* London: Christopher Helm.
2. Flood, P. R. and Deibel, D. (1998). The appendicularian house. In *The Biology of Pelagic Tunicates* (ed. Q. Bone). Oxford: Oxford University Press, pp. 105–24.
3. Collias, N. E. and Collias, E. C. (1984). *Nest Building and Bird Behaviour.* Princeton: Princeton University Press.
4. Pirk, C. W. W., Hepburn, H. R., Radloff, S. E., and Tautz, J. (2004). Honeybee combs: construction through a liquid equilibrium process? *Naturwissenschaften* 91: 350–3.
5. Read, A. T., McTeague, J. A., and Govind, C. K. (1991). Morphology and behaviour of an unusually flexible thoracic limb in the snapping shrimp, *Alpheus heterochelis. Biological Bulletin* 181: 158–68.

6. Fischer, R. and Meyer, W. (1985). Observations on rock boring by *Alpheus saxidomus* (Crustacea: Alpheidae). *Marine Biology* 89: 213–19.

Chapter 4

1. Scott Turner, J. (2000). *The Extended Organism : The Physiology of Animal-Built Structures.* Cambridge, Mass.: Harvard University Press.
2. Korb, J. and Linsenmair, K. E. (2000). Ventilation of termite mounds: new results require a new model. *Behavioural Ecology* 11: 486–94.
3. Mallon, E. B. and Franks, N. R. (2000). Ants estimate area using Buffon's needle. *Proceedings of the Royal Society of London B* 267: 765–70.
4. Frisch, K. von (1967). *The Dance Language and Orientation of Bees.* Cambridge, Mass.: Harvard University Press.
5. Jeanne, R. L. (1986) The organization of work in *Polybia occidentalis*: costs and benefits of specialization in a social wasp. *Behavioural Ecology and Sociobiology* 19: 333–41.
6. Jeanne, R. L. (1996). Regulation of nest construction behaviour in *Polybia occidentalis. Animal Behaviour* 52: 473–88.
7. Crook, J. H. (1964). Field experiments on the nest construction and repair behaviour of certain weaver birds. *Proceedings of the Zoological Society of London* 142: 217–55.
8. Grassé, P-P. (1959). La réconstruction du nid et les coordinations interindividuelles chez *Bellicositermes natalensis* et *Cubitermes* sp. La théorie de la stigmergie: Essai d'interprétation du comportment des termites constructeurs. *Insectes Sociaux* 6: 41–83.
9. Theraulaz, G. and Bonabeau, E. (1995). Modelling the collective building of complex architectures in social insects with lattice swarms. *Journal of Theoretical Biology* 177: 381–400.

Chapter 5

1. Norell, M. A., Clark, J. M., Chiappe, L. M., and Dashzeveg, D. (1995). A nesting dinosaur. *Nature* 378: 774–6.
2. Turner, A. (1989). *A Handbook to the Swallows and Martins of the World.* London: Christopher Helm.
3. Winkler and Sheldon (1993). Evolution of nest construction in swallows (Hirundinidae): a molecular phylogenetic perspective. *Proceedings of the National Academy of Sciences USA* 90: 5705–7.
4. Dawkins, R. (1982). *The Extended Phenotype*. Oxford: Freeman.
5. Dawkins, R. (1976). *The Selfish Gene.* Oxford: Oxford University Press.
6. Biron, D. G., Marché, L., Ponton, F., Loxdale, H. D., Galéotti, N., Renault, L., Joly, C., and Thomas, F. (2005). Behavioural manipulation in a grasshopper harbouring hairworm: a proteonomics approach. *Proceedings of the Royal Society B* 272 (1577): 2117–26.

7. Eberhard, W. G. (2001). Under the influence: Webs and building behaviour of *Plesiometa argyra* (Araneae, Tetragnathidae) when parasitized by *Hymenoepimecis argyraphaga* (Hymenoptera, Ichneumonidae). *The Journal of Arachnology* 29: 354–66.

Chapter 6

1. Smith, M. D. and Conway, C. J. (2007). Use of mammal manure by nesting burrowing owls: a test of four functional hypotheses. *Animal Behaviour* 73: 65–73.
2. Zschokke, S. (1996). Early stages of orb web construction in *Araneus diadematus* Clerck. *Revue Suisse de Zoologie* 2: 709–20.
3. Evans-Pritchard, E. (1967). *The Zande Trickster.* Oxford: Clarendon Press.
4. Broadley, A. and Stringer, A. N. (2001). Prey attraction by larvae of the New Zealand glowworm, *Arachnocampa luminosa* (Diptera: Mycetophilidae). *Invertebrate Biology* 120: 170–7.
5. Eberhard, W. G. (2000). Breaking the mould: behavioural variation and evolutionary innovation in *Wendilgarda* spiders (Araneae Theridiosomatidae). *Ethology, Ecology and Evolution* 12: 223–35.
6. Vollrath, F. (1992). spider's webs and silks. *Scientific American* 266 (March): 70–6.
7. Zschokke, S. (2003). Spider-web silk from the Early Cretaceous. *Nature* 424: 636–7.
8. Haynes, K. F., Gemeno, C., Yeargan, K. V., Millar, J. G., and Johnson, K. M. (2002). Aggressive chemical mimicry of moth pheromones by a bolas spider: how does this specialist predator attract more than one species of prey? *Chemoecology* 12: 99–105.
9. Bruce, M. J. (2006). Silk decorations: controversy and consensus. *Journal of Zoology* 269: 89–97.
10. Sandoval, C. P. (1994). Plasticity in web design in the spider *Parawixia bistriata*: a response to variable prey type. *Functional Ecology* 8: 701–7.
11. Heiling, A. M. and Herberstein, M. E. (1999). The role of experience in web-building spiders (Araneidae). *Animal Cognition* 2: 171–7.

Chapter 7

1. McGrew, W. C. (1992). *Chimpanzee Material Culture.* Cambridge: Cambridge University Press.
2. Washburn, S. L. (1959). Speculations on the interrelations of the history of tools and biological evolution. *Human Biology* 31: 21–31.
3. Ciochon, R. L. and Fleagle, J. G. (2006). *The Human Evolution Source Book* (2nd edn). New Jersey: Pearson, Prentice Hall.
4. Whiten, A. and Byrne, R. W. (1988). The Machiavellian intelligence hypothesis: editorial. In *Machiavellian Intelligence. Social Expertise and the*

Evolution of Intellect in Monkeys, Apes and Humans, ed. R. W. Byrne and A. Whiten. Oxford: Clarendon Press, pp. 1–9.

5. Breuer, T., Ndoundou-Hokemba, M., and Fishlock, V. (2005). First observation of tool use in wild gorillas. *Public Library of Science Biology* 3: 2041–3.
6. Hansell, M. H. (1987b). What's so special about using tools? *New Scientist* (8 January): 54–6.
7. Beck, B. B. (1980). *Animal Tool Behaviour*. New York: Garland STPM.
8. Robinson, M. H. and Robinson, B. (1971). Predatory behaviour of the ogre-faced spider *Dinopis longipes* F. Cambridge (Araneae, Dinopidae). *American Midland Naturalist* 85: 85–96.
9. Goodall, J. (1968). The behaviour of free-living chimpanzees in the Gombe Stream Reserve. *Animal Behaviour Monographs* 1: 163–311.
10. Povinelli, D. J. (2000). *Folk Physics for Apes.* Oxford: Oxford University Press.
11. Fragaszy, D., Izar, P., Visalberghi, E., Ottoni, E. B., and Olivera, M. G. de (2004). Wild capuchin monkeys (*Cebus libidinosus*) use anvils and stone pounding tools. *American Journal of Primatology* 64: 359–66.
12. Smolker, R. A., Richards, A. F., Connor, R. C., Mann, J., and Berggren, P. (1997). Sponge carrying by Indian Ocean bottle-nose dolphins: Possible tool use by a delphinid. *Ecology* 103: 454–65.
13. Lefebvre, L., Nicolakakis, N, and Boire, D. (2002). Tools and brains in birds. *Behaviour* 139: 939–73.
14. Hunt, G .R. (1996). Manufacture and use of hook tools by New Caledonian crows. *Nature* 379: 249–51.
15. Hunt, G. R. (2000). Human-like, population-level specialization in the manufacture of pandanus tools by the New Caledonian crows *Corvus moneduloides. Proceedings of the Royal Society of London B* 267: 403–13.
16. Weir, A. A. S., Chappell, J., and Kacelnik, A. (2002). Shaping of hooks in New Caledonian crows. *Science* 297: 981.
17. Weir, A. A. S. and Kacelnik, A. (2006). A New Caledonian crow (*Corvus moneduloides*) creatively re-designs tools by bending or unbending aluminium strips. *Animal Cognition* 9: 317–34.
18. Weir, A. A. S. and Kacelnik, A. (2005). Behavioural ecology: tool manufacture by naïve juvenile crows. *Nature* 433: 121.
19. Tebbich, S., Taborsky, M., Fessl, B., and Blomqvist, D. (2001). Do woodpecker finches acquire tool-use by social learning? *Proceedings of the Royal Society of London B* 268: 2189–93.

Chapter 8

1. Frith, C. B. and Frith, D. W. (2004). *The Bowerbirds.* Oxford: Oxford University Press.

2. Uy, J. A. C. and Borgia, G. (2000). Sexual selection drives rapid divergence in bowerbird display traits. *Evolution* 54: 273–8.
3. Madden, J. (2002). Bower decorations attract females but provoke other male spotted bowerbirds: bower owners resolve this trade-off. *Proceedings of the Royal Society of London B* 269: 1347–51.
4. Morris, D. (1962). *The Biology of Art.* London: Methuen.
5. Wilson, E. O. (1975). *Sociobiology. The New Synthesis.* Cambridge, Mass.: Harvard University Press.
6. Dawkins, R. (1976). *The Selfish Gene.* Oxford: Oxford University Press.
7. Dissanayake, E. (1988). *What is Art for?* Seattle: University of Washington Press.
8. Dissanayake, E. (1995). *Homo Aestheticus.* New York: The Free Press.
9. McGrew, W. C. and Marchant, L. F. (1998). Chimpanzee wears a knotted skin 'necklace'. *Pan African News* 5: 8.
10. Rachter (1984). *The Policeman's Beard is Half Constructed. Computer Prose and Poetry by Rachter.* New York: Warner Books.
11. Miller, G. (2000). *The Mating Mind.* London: Heinemann.
12. Wojcieszek, J. M., Nicholls, J. A., Marshall, N. J., and Goldizen, A. W. (2006). Theft of bower decorations among male satin bowerbirds (*Ptilonorhynchus violaceus*): why are some decorations more popular than others? *Emu* 106: 175–80.
13. Bravery, B. D., Nicholls, J. A., and Goldizen, A. W. (2006). Patterns of painting in satin bowerbirds *Ptilonorhynchus violaceus* and males' responses to changes in their paint. *Journal of Avian Biology* 37: 77–83.
14. Patricelli, G. L., Coleman, S. W., and Borgia, G. (2006). Male satin bowerbirds, *Ptilonorhynchus violaceus*, adjust their display intensity in response to female startling: an experiment with robotic females. *Animal Behaviour* 71: 49–59.
15. Endler, J. A. and Day, L. B. (2006). Ornament colour selection, visual contrast and the shape of colour preference functions in great bowerbirds, *Chlamydera nuchalis. Animal Behaviour* 72: 1405–16.
16. Day, L. B., Westcott, D. A., and Olster, D. H. (2004). Evolution of bower complexity and cerebellum size in bowerbirds. *Behaviour, Brain and Evolution* 66: 62–72.
17. Duncan, I. J .H. (2006). The changing concept of animal sentience. *Applied Animal Behaviour Science* 100: 11–19.
18. Kohn, M. (1999). *As We Know It.* London: Granta Books.
19. Stringer, C. and Andrews, P. (2005). *The Complete World of Human Evolution.* London; New York: Thames and Hudson.

Index